T0215820

# Freiburger Empirische Forschung in der Mathematikdidaktik

Die Freiburger Arbeitsgruppe am Institut für Mathematische Bildung (IMBF) verfolgt in ihrem Forschungsprogramm das Ziel, zur empirischen Fundierung der Mathematikdidaktik als Wissenschaft des Lernens und Lehrens von Mathematik beizutragen. In enger Vernetzung innerhalb der Disziplin und mit Bezugsdisziplinen wie der Pädagogischen Psychologie oder den Erziehungswissenschaften sowie charakterisiert durch eine integrative Forschungsmethodik sehen wir Forschung und Entwicklung stets im Zusammenhang mit der Qualifizierung von wissenschaftlichem Nachwuchs. Die vorliegende Reihe soll regelmäßig über die hierbei entstehenden Forschungsergebnisse berichten.



More information about this series at http://www.springer.com/series/10531

Benjamin Rott

# Epistemological Beliefs and Critical Thinking in Mathematics

## Qualitative and Quantitative Studies with Pre-Service Teachers

Benjamin Rott
Mathematikdidaktik
Universität zu Köln
Köln, Germany

ISSN 2193-8164 ISSN 2193-8172 (electronic)
Freiburger Empirische Forschung in der Mathematikdidaktik
ISBN 978-3-658-33538-0 ISBN 978-3-658-33539-7 (eBook)
https://doi.org/10.1007/978-3-658-33539-7



Responsible Editor: Marija Kojic
This Springer Spektrum imprint is published by the registered company Springer Fachmedien Wiesbaden GmbH part of Springer Nature.
The registered company address is: Abraham-Lincoln-Str. 46, 65189 Wiesbaden, Germany

# Foreword

Research on epistemological beliefs—even if it is diverse—has consistently been based on the premise that naïve epistemological beliefs at an early stage go hand in hand with the conviction that (scientific) knowledge is safe, while sophisticated beliefs emphasize the uncertainty of scientific knowledge. This one-dimensional view is, according to the starting point of Benjamin Rott's research, untenable: In the domain of mathematics, experts with a high degree of sophistication can at the same time emphasize the certainty of mathematical knowledge. Here a distinction must be made between beliefs as dispositions and contextually varying argumentations.

As plausible as this statement is—it has not been pursued in research to date, with a few exceptions, and empirical evidence is lacking. In his studies, Benjamin Rott shows convincingly and systematically how such an empirical foundation can be established. Survey instruments are developed and validated (using people with varying degrees of expertise), which make it possible to make the phenomena in question visible. Existing instruments (especially closed questionnaires) are further developed and supplemented by semi-open interview forms and rating procedures. In addition, with reference to research on Critical Thinking, a performance test is conceived in such a way that it can record the sophistication of mathematical thinking—largely independently of specialist knowledge. On this basis, the postulated independence of sophistication and security conviction is demonstrated in various studies: In structured interview studies as well as in cross-sectional and longitudinal surveys with students. Benjamin Rott shows how the intelligent use of qualitative and quantitative methods can inspire and support findings from each perspective.

Benjamin Rott's research has by now gained considerable attention in the mathematics education research. The publication of a book which integrates, reflects and extends existing research can make additional contributions to theory formation and research strategies in the area of mathematical beliefs.

Freiburg
August 2020

Timo Leuders

# Preface

Nowadays, results from research projects are often scattered across several publications. Results from pilot studies are published in different journals than results from main studies; some data are published only in conference proceedings and some data are not published at all. Additionally, journal publications are often subject to strict word or character limitations, which leads to shortened presentations. For example, usually only coefficients for correlations and interrater agreements are given and according tables that can transport additional information have to be shortened.

All of the above is true at least for the research project LeScEd (Learning the Science of Education). This book gives me the great opportunity to collect and bring together all the research results from sub-project #3 of this project—and to literally press them between the covers of a book. In this sense, the Chapters 2, 3, 4, 5, and 6 of this book are, therefore, mostly reprints of previously published articles that have been slightly adapted and extended (e.g., including tables showing interrater agreement that had to be cut before publication in journals; or standardizing terminology like "epistemic beliefs"/"epistemological beliefs"). Chapters 1 and 7 provide an introduction and a framework for the study in general.

Finally, I would like to thank the persons that I collaborated with during my work at this project, which are most notably Elmar Stahl, Jana Groß Ophoff, and especially Timo Leuders.

Cologne
August 2020

Benjamin Rott

# Abstract

The research in this book deals with epistemological beliefs, i.e. beliefs on the nature of knowledge, its limits, sources, and justification. More precisely, studies dealing with epistemological beliefs in the domain of mathematics are presented.

In this research tradition, usually closed instruments (most often Likert scale, self-report questionnaires) are used to measure beliefs. Participants are asked to state their agreement to several knowledge claims. Positions regarding certain, unchanging knowledge are associated with naïve, unreflected beliefs; in line with this, positions regarding uncertain, tentative knowledge are associated with sophisticated beliefs. However, there is an ongoing discussion about the reliability and validity of such instruments, indicating problems with social desirability, the interpretation of items, domain-specificity of beliefs, etc.

Therefore, as part of the study at hand, a new instrument was developed to specifically address mathematics-related epistemological beliefs as mathematics is a scientific domain with ways of generating and justifying knowledge unlike any other domain.

Results of an initial qualitative interview study with pre- and in-service mathematics teachers as well as mathematicians and mathematics educators show the special status of epistemological beliefs with regard to mathematics. For this domain, the association of belief position to belief sophistication seems not to be true. For example, interviewees held belief positions about certainty of knowledge with very convincing (i.e., sophisticated) arguments, whereas others held belief positions about uncertainty of knowledge, arguing in a very unreflected way.

These findings regarding belief positions and corresponding argumentations were further explored in quantitative studies (with $n = 147$, $n = 463$, and $n = 439$ pre-service teachers) using an open-ended questionnaire that was built on the experiences of the interview study. The results of these studies all confirm statistical independence of belief position and belief argumentation for the belief dimensions "certainty" and "justification of mathematical knowledge".

Finally, a test for mathematical critical thinking (CT)—as a component of students' competencies—was developed. This test consists of several tasks that are supposedly easy to solve but whose results need to be checked, this way measuring a disposition to reflect upon results. Using it in all three quantitative studies revealed that belief position is not correlated to CT-test results, whereas sophisticated belief argumentations are significantly correlated to high CT-test scores. These correlative results further suggest to differentiate between belief positions and argumentation and not associate belief sophistication with the former.

# Zusammenfassung

Die Forschung in diesem Buch beschäftigt sich mit epistemologischen Überzeugungen, d. h. Beliefs über die Natur menschlichen Wissens, seine Grenzen, Quellen und Rechtfertigungen. Konkret werden Studien vorgestellt, die sich mit epistemologischen Überzeugungen in Bezug auf die Mathematik befassen.

In dieser Forschungstradition werden in der Regel geschlossene Instrumente (meist Likert-Skalen, Selbstauskunft-Fragebögen) zur Messung von Beliefs verwendet. Teilnehmende sollen ihre Zustimmung zu Aussagen angeben. Positionen in Bezug auf sicheres, unveränderliches Wissen werden mit naiven, unreflektierten Beliefs assoziiert; passend dazu werden Positionen in Bezug auf unsicheres, vorläufiges Wissen mit reflektierten Beliefs assoziiert. Diskutiert wird jedoch die Reliabilität und Validität solcher Instrumente, insb. in Bezug auf Probleme mit sozialer Erwünschtheit, der Interpretation von Items, der Domänenspezifität usw.

Daher wurde im Rahmen der vorliegenden Studie ein neues Instrument entwickelt, das sich speziell mit mathematikbezogenen epistemologischen Beliefs befasst, da die Mathematik eine wissenschaftliche Domäne ist, die Wege zur Generierung und Rechtfertigung von Wissen besitzt wie keine andere.

Die Ergebnisse einer ersten, qualitativen Interviewstudie mit Studierenden, Lehrer*innen, Mathematiker*innen und -Didaktiker*innen zeigen den besonderen Stellenwert epistemologischer Überzeugungen in Bezug auf die Mathematik. Hier scheint die Assoziation von Belief-Position und -Reflektiertheit nicht zu passen. Einige Interviewte vertraten z. B. die Position „Wissen ist sicher“ mit sehr überzeugenden (reflektierten) Argumenten, während andere die Position „Wissen ist unsicher“ vertraten und dabei sehr unreflektiert argumentierten.

Die Ergebnisse in Bezug auf Belief-Positionen und zugehörige Argumentationen wurden in quantitativen Studien (mit $n = 147$, $n = 463$ und $n = 439$

Lehramtsstudierenden) – auf Basis der Interviewstudie – mit einem offenen Fragebogen weiter untersucht. Die Ergebnisse aller drei Studien bestätigen eine statistische Unabhängigkeit von Belief-Positionen und -Argumentationen für die Belief-Dimensionen „Sicherheit" und „Rechtfertigung mathematischen Wissens".

Schließlich wurde ein Test für mathematisch-kritisches Denken (CT) – als Bestandteil der Kompetenzen von Studierenden – entwickelt. Dieser Test besteht aus mehreren, vermeintlich leicht zu lösenden Aufgaben, deren Ergebnisse jedoch überprüft werden müssen, um so eine Disposition zur Ergebnisreflexion zu messen. In allen drei quantitativen Studien zeigte sich, dass die Belief-Position nicht mit den Ergebnissen des CT-Tests korreliert ist, wohingegen reflektierte Belief-Argumentationen signifikant mit hohen CT-Testergebnissen korreliert sind. Auch diese korrelativen Ergebnisse legen also nahe, zwischen Belief-Positionen und -Argumentation zu unterscheiden und Reflektiertheit nicht ausschließlich mit der Position in Verbindung zu bringen.

# Contents

# Introduction 1

In this book, several years of research dealing with the topic "epistemological beliefs" are summarized. Specifically, research focusing on the development and evaluation of an instrument to measure such beliefs are presented, as traditional instruments are often criticized for a lack of validity and reliability.

The chapters of this book can be read more or less independent of each other as they are based on previously published articles. This first chapter (together with Chap. 7) is used to provide an overarching structure for the entire book. In the first section of this chapter, the motivation of addressing the topic "epistemological beliefs" is explained. This is followed by a general introduction into and a theoretical framework of this topic. Finally, an introduction to the research project that initiated the scientific studies presented in this book is given in the third section of this chapter. This goes along with an overview of the contents of the book as well as a summary of the individual publications.

## 1.1 Motivation

"Epistemological beliefs" is probably a term with which even many scientists are not very familiar; not to mention the fact that the majority of non-scientists will most likely have never heard of it. Yet, the concept of epistemological beliefs is highly relevant not only for research but also for teaching and learning as well as everyday situations.

Epistemological beliefs (a more precise definition follows in Sect. 1.2) are views of the nature of our knowledge, addressing questions like the following ones: "Where do 'facts' come from?", "How do scientists come up with research



B. Rott, *Epistemological Beliefs and Critical Thinking in Mathematics*, Freiburger Empirische Forschung in der Mathematikdidaktik,
https://doi.org/10.1007/978-3-658-33539-7_1

results?", etc. How do views on such questions affect our thoughts and even our actions? Three brief examples shall illustrate the importance of this topic.

In the summer of 2020, the COVID-19 pandemic determines our lives. Arthur informs himself about the use of wearing cloth face masks. Do they protect him? Do they protect others? Or are they an instrument of suppression, to muzzle the public? Whom should he belief? Should he trust in scientists like Anthony Fauci or Christian Drosten who now strongly recommend wearing masks in public, but might have changed their position regarding the effectiveness of such masks within the last months? Or should he believe in politicians like Donald Trump or even in conspiracy theorists who disapprove of or even call upon resisting against using face masks? Without understanding how scientists work and why they might change their positions with new insights, Arthur might believe in people who are against wearing face masks for ideological reasons.

Bianca is a high school student who, in physics classes, perceived scientific experiments in the following way: The teacher demonstrates an—often astonishing—phenomenon; thereafter, the teacher, together with the class, develops a physical theory that explains the experiment. For Bianca, physical experiments are random observations, but not goal-oriented actions. Confronted with an experiment like the Rutherford scattering experiment, in which alpha particles are fired against gold foil to explore the nature of atomic nuclei, Bianca is puzzled. How can anyone observe such a strange phenomenon? Without appropriate views of scientific approaches—in this case: hypothesis-driven experimentation—she is going to have a hard time understanding physics.

Christina is a mathematics teacher. In her own school days as well as in her university studies to become a teacher, she experienced mathematics in a special way: Mathematical facts or procedures were explained to her by figures of authority like teachers or professors. Mathematical rules or theorems were always justified (or even proven) in a deductive way, well-ordered in textbooks or scripts. From her experiences, Christina concludes that this must be the way mathematics and mathematicians work. For her teaching, therefore, nothing else is possible but reproducing this way of knowledge transfer; she does not even consider problem-oriented teaching and discovery learning.

Hopefully, these three examples help in illustrating the importance of research on the topic of epistemological beliefs. Throughout the book, it will be shown that even though there is a long tradition in this line of research, there are problems with the reliability and validity measuring such beliefs. Especially closed-ended, self-report instruments, which are prominently used in research from psychology as well as from mathematics education (e.g., COACTIV or TEDS-M) should be reflected upon critically. This motivates the development and evaluation of an open-ended, quantitative instrument to measure epistemological beliefs in the domain of mathematics, which is the main focus of this book.

## 1.2 Research on the Topic of Epistemological Beliefs

Empirical studies on epistemological beliefs is a field of research that originated in psychology and educational research that deals with questions like "What do people think about how our knowledge comes into being or is justified?" (cf. Hofer and Pintrich 1997). Nowadays, this research topic has also been taken up by subject-specific education research like physics (or more broadly science) education or mathematics education. Before research on epistemological beliefs is addressed specifically, its components—epistemology and beliefs—are introduced separately.

### 1.2.1 Epistemological Beliefs—a Clarification of Terms

**Epistemology** (Greek: επιστεμε, epistēmē = knowledge; λογος, logos = theory, logical discourse) is a branch of philosophy dealing with the nature and justification of human knowledge; addressed are questions like "What is knowledge? How is knowledge being generated and how certain is it?" (cf. Arner 1972 and Butchvarov 1970). Arner (1972) distinguishes three major areas: Questions related to the limits of human knowledge, to the sources from which knowledge is generated (such as inspiration, myth, science, experience), and to the nature of human knowledge (e.g., "When can a statement be considered true?").

In later chapters of this book, questions regarding the certainty and justification of knowledge will be presented and discussed in the context of the philosophy of mathematics.

Research on **beliefs** (Proto-Germanic *laubô*, German "Glaube" = faith, belief) deals with what philosophers characterize as a "propositional attitude", that is "the mental state of having some attitude, stance, take, or opinion about a proposition or about the potential state of affairs in which that proposition is true" (Schwitzgebel 2015, 2nd par.). Similar definitions are used in research from psychology and mathematics education, for example by Goldin or Schoenfeld.

> Beliefs are defined to be multiply-encoded cognitive/affective configurations, to which the holder attributes some kind of truth value (e.g., empirical truth, validity, or applicability). (Goldin 2002, p. 59)

> Beliefs – to be interpreted as 'an individual's understandings and feelings that shape the ways that the individual conceptualizes and engages in mathematical behavior' [...]. (Schoenfeld 1992, p. 358)

Philipp (2007, p. 259) characterizes beliefs as "psychologically held understandings, premises, or propositions about the world that are thought to be true." Beliefs, Philipp (ibid.) continues, "might be thought of as lenses that affect one's view of some aspect of the world or as a disposition toward action." Fishbein and Ajzen (1975) also mention an effect of beliefs on acting and say that beliefs are assumed to control, but not determine, our behavior and thought processes.

Philipp points out that beliefs can refer to different subjects like learning mathematics, one's own abilities, or the truth of scientific theories like evolution theory. He further states that they are often organized in clusters around particular ideas or objects, called belief systems, a concept introduced by Green (1971).

Many researchers point out that the term "belief" is not clearly defined (e.g., Furinghetti and Pehkonen 2002; Hart 1989; Philipp 2007; Thompson 1992). Furinghetti summarizes possible reasons for these terminological differences:

> Partly this disagreement is due to mere linguistic questions since in some cases the translation of terms from one language to another may modify the original meaning. Another cause of disagreement is the different sense, epistemic or attitudinal, that may be ascribed to the term 'belief', according to which the perspective of the discussion changes considerably. Also the context in which a given research is set may change the assumptions of the various authors. In addition, sometimes I have observed that the various authors do not state explicitly the hypotheses on which their work relies. (Furinghetti 1998, p. 24)

Regarding the sense or the perspective that Furinghetti speaks about, two main positions emerge from the literature. On the one hand, beliefs can be conceptualized in an *epistemological or ontological* context (see Fig. 1.1, top), differentiating between objective (or true) knowledge and subjective (not necessarily justified) knowledge, that is beliefs (cf. Murphy and Mason 2006; Pehkonen 1999). Sometimes, under this perspective, knowledge is characterized as "true belief" in a Platonic sense, which is not accessible to humans (see Philipp 2007, pp. 266 ff. for a further discussion). On the other hand, in the *attitudinal* context (see Fig. 1.1, bottom), beliefs are seen as a part of the concept of affect with emotions, attitudes, and beliefs—in this order—being less or more cognitive, less or more stable, and being felt more or less intense (Philipp 2007; McLeod 1989; see also Furinghetti and Pehkonen 2002; Schoenfeld 1985). In the research project at hand, neither

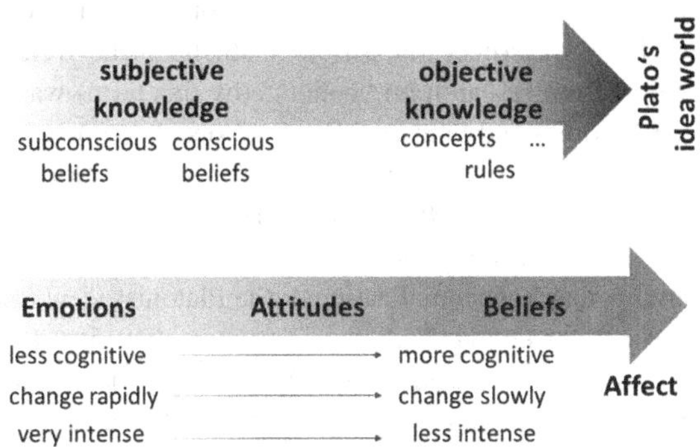


**Fig. 1.1** Two different conceptualizations of "beliefs"

approach to conceptualize beliefs is regarded as false or "more correct" than the other; more details are given in Chap. 4.

As stated above, beliefs refer to different subjects; beliefs regarding questions of epistemology are called **epistemological beliefs**. Starting with Piaget (1950) and Perry (1970), psychologists investigate the extent to which students hold beliefs regarding the nature of knowledge and knowing and whether or in which way these beliefs affect or are correlated with behavior, learning gains, etc. (cf. Briell et al. 2011; Hofer and Pintrich 1997). In this context, in the literature from psychology and mathematics education, different terms are used more or less synonymously. Amongst others, these terms are *epistemological beliefs*, *epistemic beliefs*, and *personal epistemology* in psychology (see Briell et al. 2011 for more terms); *mathematical beliefs*, *beliefs about (the nature of) mathematics*, or *mathematical world views* in mathematics education; and in science education, the term *(beliefs about the) nature of science* is widely used (cf. Höttecke 2001; Neumann and Kremer 2013).

The term personal epistemology was specified by Hofer and Pintrich (1997) and is often used in connection with the development of corresponding beliefs (cf. Briell et al. 2011, p. 7). Briell et al. (ibid., p. 14 f.) discuss the distinction between epistemic and epistemological beliefs. As an adjective, epistemic means that corresponding beliefs are knowledge in the literal sense of the word (i.e., justified, true beliefs), whereas epistemological beliefs refer to the theory of knowledge. They therefore argue for the use of the term epistemological to identify respective beliefs (which is used in this book).

In mathematics education, Schoenfeld (1985) coined the term "mathematical world view" (used as the German translation "mathematische Weltbilder" by Grigutsch et al. 1998) before research on "beliefs" (by this term) was established in this community (cf. Törner and Pehkonen 1999). Nowadays, "world views" is still used, often synonymously to "mathematical beliefs". Both terms regularly refer to beliefs described by Dionné (1984), Ernest (1989), and Grigutsch et al. (1998) with the following or similar terms (see Rott 2018, 2020 for comparisons of these three more or less similar conceptualizations): the Platonist view (mathematics as a unified body of certain knowledge), the instrumentalist view (mathematics as a collection of rules and facts), and the problem-solving view (mathematics as a dynamic, expanding field of human creation and invention).

In this book, the term "epistemological beliefs" is used. If necessary, the term will be specified as "epistemological beliefs regarding mathematics" (or similar), but the psychological approach of conceptualizing epistemological beliefs is used instead of the mathematical conceptualization of "views".

### 1.2.2 History of Research on Epistemological Beliefs

The following overview of research on epistemological beliefs and on developmental models is based on Hofer (2001) and, in particular, on Hofer and Pintrich (1997), who present a detailed description and discussion of these models. Research on epistemological beliefs and especially on developmental models regarding this topic begins with Piaget (1950, 1974) and his theory on the development of intelligence in stages; all other models expand on Piaget's ideas.

In the 1950s and 1960s, Perry (1968, 1970) conducted two long-term studies with students and developed a first model for the development of epistemological beliefs. Based on interviews, nine development steps were originally identified and validated in the second study. Perry's development steps are typically grouped into four consecutive categories: *dualism* (absolutist, right-wrong, authority-believing), *multiplicity* (uncertainties and diversities are partially accepted; truth is still assumed to be ascertainable), *relativism* (knowledge is relative, everyone can generate knowledge themselves), and *commitment within relativism* (responsibility, engagement in relativistic positions). Almost all subsequent studies refer to and build on Perry's research.

Based on the criticism that Perry had only researched male subjects and only elite students, Belenky et al. (1986) interviewed women with different backgrounds. In their study, called *Women's Ways of Knowing*, they introduced a model with five levels, some of which coincide with Perry's: *silence* (passive, believing authority), *received knowing* (dualistic, knowledge can be reproduced), *subjective knowledge* (dualistic, source of truth lies in the persons themselves), *procedural knowledge* (reasoned reflection, systematic analysis), and *constructed knowledge* (knowledge and truth are constructed and contextual); sometimes dividing *procedural knowledge* into *connected* and *separate knowing*.

For her *Epistemological Reflection Model*, Baxter Magolda (1992) carried out a five-year long-term study with 50 men and 51 women and identified four different levels. The levels are *absolute* (knowledge is certain, believing in authority), *transitional* (recognition that authorities are not omniscient and knowledge can be uncertain), *independent* (authorities are questioned, one's own opinion is accepted as valid) and *contextual* (individual perspective through assessment of facts in context).

King and Kitchener (1994) have established the *Reflective Judgment Model* with seven levels of how individuals perceive and deal with ill-structured problems in long-term interview studies. These levels are summarized in the following three levels: *pre-reflective* (learners can take that there are no unsolvable problems), *quasi-reflective* (knowledge is not certain) and *reflective* (knowledge is actively constructed and contextually understood).

Kuhn (1991) conducted a particularly wide-ranging interview study with subjects from four age groups—from teenagers to over sixty years old. He examined argumentative reasoning for everyday, ill-structured problems. Epistemological beliefs are categorized into three levels: *absolutist* (knowledge is secure and absolute, believers in authority), *multiplist* (skepticism about expertise, authorities are not recognized), or *evaluativist* (there is no secure knowledge, experts are accepted as better informed).

Hofer summarizes these five very influential models as follows:

> These models share interactionist, constructivist assumptions and sketch similar trajectories of development. The path of epistemological development begins with an objectivist, dualistic view of knowledge, followed by a multiplistic stance, as individuals begin to allow for uncertainty. Typically, a period of extreme subjectivity is followed by the ability to acknowledge the relative merits of different points of view and to begin to distinguish the role that evidence plays in supporting one's position. In the final stage, knowledge is actively constructed by the knower, knowledge and truth are evolving, and knowing is coordinated with justification. This culminating perspective has been variously labeled commitment within relativism (Perry 1970), reflective thinking (King and Kitchener 1994), constructed knowledge (Belenky et al. 1986), contextual knowing (Baxter Magolda 1992), or evaluativism (Kuhn 1991). (Hofer 2001, p. 359)

In contrast to the development models presented above, which are all one-dimensional, i.e. have stages building on each other, Schommer (1990, 1993) proposed a multidimensional construct. Based on works by Perry (1968), Schoenfeld (1985) and others, she worked out five dimensions (*structure*, *certainty* and *source of knowledge*, as well as *control* and *speed of knowledge acquisition*). She then tested this model in a study using a questionnaire she had developed. By means of factor analysis she was able to empirically confirm four of these dimensions (cf. Hofer and Pintrich 1997, p. 106 f.; Schommer 1998, p. 130): *fixed ability* (intelligence as something innate and fixed or as something trainable or intermediate); *quick learning* (learning between the extremes very quickly or not at all and step by step, gradually); *simple knowledge* (knowledge between isolated facts and highly networked concepts); as well as *certain knowledge* (How safe/insecure is knowledge?). The fifth dimension, *source of knowledge*, could not be proven as an independent factor using the original questionnaire.[1]

Hofer and Pintrich (1997) and Hofer (2000) finally present their own, often-quoted categorization of the concept of epistemological beliefs. For this, they removed dimensions that refer exclusively to educational experiences and learning instead of knowledge; they also criticize the inclusion of Schommer's beliefs about the ability to learn and speed of learning. The Hofer and Pintrich model has two main dimensions: *nature of knowledge* and *nature or process of knowing* with two sub-dimensions each.

The use of (one- or multi-dimensional) linear developmental models strongly suggests the following progression of learners' epistemological beliefs from naïve to sophisticated. In the beginning of this development, there are beliefs of certainty, of an objective truth, and in authorities; then, those beliefs slowly allow for uncertainties, constructed knowledge, and authorities are questioned. I repeat a part of Hofer's summary from above:

> The path of epistemological development begins with an objectivist, dualistic view of knowledge, followed by a multiplistic stance, as individuals begin to allow for uncertainty. Typically, a period of extreme subjectivity is followed by the ability to acknowledge the relative merits of different points of view and to begin to distinguish the role that evidence plays in supporting one's position. In the final stage, knowledge is actively constructed by the knower, knowledge and truth are evolving, and knowing is coordinated with justification. (Hofer 2001, p. 359)

[1] The dimensions proposed by Schommer are not universally accepted but are discussed critically in the community (cf. Briell et al. 2011). For example, in two empirical studies, Clarebout et al. (2001) were not able to reproduce the factor structure proposed by Schommer with a translated version of her questionnaire. In fact, in their second study, Clarebout et al. even were not able to reproduce the factor structure of their first study.

A similar transition from naïve to sophisticated beliefs is described by Hofer and Pintrich with regards to the dimension of certainty of knowledge.

> Certainty of knowledge. The degree to which one sees knowledge as fixed or more fluid appears throughout the research, again with developmentalists likely to see this as a continuum that changes over time, moving from a fixed to a more fluid view. **At lower levels, absolute truth exists with certainty. At higher levels, knowledge is tentative and evolving.** […] (Hofer and Pintrich, 1997, p. 119; accentuation by BR)

Comparable assumptions can be found in mathematics education.

> The majority of research that has examined students' beliefs about mathematics suggests that students at all levels hold **nonavailing beliefs. In general, when asked about the certainty of mathematical knowledge, students believe that knowledge is unchanging.** The use and existence of mathematics proofs support this notion, and students believe the goal in mathematics problem solving is to find the right answer. […] (Muis 2004, p. 330; accentuation by BR)

Such conceptualizations lead to a strong link between belief positions (most notably beliefs in the certainty of knowledge) and the assumption of the degree of sophistication of a person's beliefs.

Especially with instruments like (closed) questionnaires (see Table 1.1), belief positions are used to determine whether a person holds naïve or sophisticated beliefs. In this book, for the domain of mathematics, this link or association is questioned. It will be shown that strong beliefs in the certainty of mathematical knowledge do not necessarily indicate naïve epistemological beliefs.

In the beginning of research on epistemological beliefs (see above), extensive interviews were used to gain access to students' beliefs. Ever since Schommer's contributions to this line of research, most studies use closed (Likert-scale) questionnaires to measure epistemological beliefs (cf. Clarebout et al. 2001). Sample items from such questionnaires are listed in Table 1.1.

As stated above, in such questionnaires, belief positions (e.g., regarding the certainty of knowledge) are used to determine the stage of development of the participants' epistemological beliefs. For example, Wegner et al. (2012) directly refer to beliefs in stability and completeness of knowledge as naïve, whereas beliefs in dynamic, instable, open, and incomplete knowledge are regarded as experienced or sophisticated.

**Table 1.1** Sample items from closed questionnaires measuring (epistemological) beliefs; participants state agreement to each item on a Likert-scale

| Sample Item | Source |
|---|---|
| 1. Most things worth knowing are easy to understand | Schraw et al. (2002) |
| 2. What is true is a matter of opinion | |
| 10. Too many theories just complicate things | |
| 12. Instructors should focus on facts instead of theories | |
| 15. If you don't learn something quickly, you won't ever learn it | |
| 17. Things are simpler than most professors would have you believe | |
| 18. If two people are arguing about something, at least one of them must be wrong | |
| 21. Science is easy to understand because it contains so many facts | |
| 22. The more you know about a topic, the more there is to know | |
| 27. Working on a problem with no quick solution is a waste of time | |
| 1.1 Math problems that take a long time don't bother me | Kloosterman and Stage (1992) |
| 1.4 If I can't do a math problem in a few minutes, I probably can't do it at all | |
| 2.4 Any word problem can be solved if you know the right steps to follow | |
| 3.2 A person who doesn't understand why an answer to a math problem is correct hasn't really solved the problem | |
| 3.5 Getting a right answer in math is more important than understanding why the answer works | |
| 4.1 A person who can't solve word problems really can't do math | |

(continued)

**Table 1.1** (continued)

| Sample Item | Source |
|---|---|
| 4.4 Learning computational skills is more important than learning to solve word problems | |
| 5.4 Ability in math increases when on studies hard | |
| 2. The only thing that is certain is uncertainty itself | Schommer (1990) retrieved from Bierman (2008, pp. 309 ff.) |
| 6. You can believe almost everything you read | |
| 12. If scientists try hard enough, they can find the truth to almost anything | |
| 21. Scientists can ultimately get to the truth | |
| 31. Being a good student generally involves memorizing facts | |
| 32. Wisdom is not knowing the answers, but knowing how to find the answers | |
| 34. Truth is unchanging | |
| 40. Sometimes you just have to accept answers from a teacher even though you don't understand them | |
| 57. An expert is someone who has a special gift in some area | |
| 59. The best thing about science courses is that most problems have only one right answer | |
| 61. Today's facts may be tomorrow's fiction | |

### 1.2.3 Relevance of This Topic

It is generally assumed that epistemological beliefs play an important role, both in everyday life and in learning processes of all kinds; starting with **theoretical considerations**, there are several reasons that imply the relevance of this topic.

In order to receive information meaningfully from newspapers and other media or—more generally—to actively participate in our modern knowledge society, an adequate understanding of the nature of (natural) sciences is necessary (cf. Elen et al. 2011, p. 1).

In the acquisition of knowledge—be it at school, university, or other occasions—learners are confronted with (scientific) information presented to them by teachers, in books, films, or other media. It is then almost always a matter of checking this information with regard to its sources and its truth content, or of assessing its plausibility (cf. Stahl 2011, p. 38).

Furthermore, epistemological beliefs can have subtle but comprehensive effects on learning processes. For example, on the one hand, for a person who believes that knowledge is *structured* in isolated pieces, learning mostly consists of memorizing lists; with the result that this person can quote facts well but apply knowledge poorly. Someone who perceives knowledge as networked, on the other hand, is more likely to learn comprehensively; such a person is likely to be able to apply knowledge well, but will not be quite as fast as the first person in reproducing facts (cf. Schommer 1998, p. 132; Tsai 1998a, b). With regard to *simplicity* and *certainty of knowledge*, someone who understands knowledge in this way could look for "the one" correct method in mathematical problems, whereas someone who understands knowledge as provisional and complex probably looks for different approaches and solutions (Schommer 1998, p. 133 f.).

For these reasons, the formation of sophisticated epistemological beliefs is considered an educational goal in itself (cf. Bromme and Kienhues 2008; Trautwein and Lüdtke 2007, p. 362).

Of course, this also applies in particular to teachers who make an appropriate selection as to which information they present and how they present it (and which information they do not pass on to the learners) (cf. Wegner and Nückles 2011, p. 173).

Regarding **empirical reasons** for the relevance of research on epistemological beliefs, studies have shown that epistemological beliefs influence the learning process, are predictors of the success of learning processes (for an overview see e.g. Hofer and Pintrich 1997; Schommer 1998), and are significantly related to academic achievement.

For example, Schommer (1990) found a correlation between students' education and their beliefs: "The more classes the students [in her study] had completed in higher education, the more likely they were to believe knowledge is tentative." (ibid., p. 501) In 1993, Schommer recorded and correlated the epistemological beliefs of 869 high school students with the help of the questionnaire she had developed and the grade point averages (GPA). She was able to demonstrate significant correlations between sophisticated beliefs and good grades:

> The less that students believed in quick learning (r = −0.26), simple knowledge (r = −0.20), certain knowledge (r = −0.12), and fixed ability (r = −0.15), the better were the GPAs that they earned. [...]
>
> The most intriguing finding is that epistemological beliefs predicted GPA. A conservative interpretation of the results is that at least one epistemological belief – belief in quick learning – predicts academic performance, even after general intelligence has been taken into consideration. (Schommer, 1993, pp. 409 f.)

The same applies to Hofer (2000), who used questionnaires (revised versions of Schommer's questionnaire) to collect the epistemological beliefs of more than 300 first-year college students of a psychology course and correlated them with their grades in psychology, the natural sciences, and their grade point averages for the semester. In general, there were negative correlations between naive epistemological beliefs and good grades, with the aspect "safety and simplicity of knowledge" in particular showing high and significant correlations.

Trautwein and Lüdtke (2007) recorded the epistemological beliefs of 2854 high school graduates from 90 grammar schools (Gymnasien). In addition to the overall grade at the Abitur, epistemological beliefs of the dimension certainty of knowledge were surveyed, using a questionnaire based on the instruments of Hofer (2000) and Schommer (1990), as well as items they created themselves. Structural equation models were used to demonstrate a significant negative relationship between less sophisticated beliefs and the Abitur grade.

> Taken together, in line with our expectations, we found certainty beliefs to negatively predict school achievement, even when important other variables were controlled. Certainty beliefs partly mediated the impact of cognitive abilities, gender, and cultural capital on school achievement. Given that access to highly valued fields of study is competitive in Germany, the negative effect of certainty beliefs on final school grades was by no means negligible. (Trautwein and Lüdtke, 2007, p. 359)

## 1.3 The Research Project LeScEd

Research orientation is a key characteristic of higher education and university education (cf. Tremp and Futter 2012). It is represented in normative frameworks for educational studies such as teacher education (e.g., KMK 2004). Research orientation is characterized as the competence to receive and understand scientific knowledge ("engagement with research") and in addition to think and work scientifically ("engagement in research") (cf. Borg 2010). A development of these competencies is seen as essential to prepare pedagogical and educational professions in understanding science and science communication (cf. Groß Ophoff and Rott 2017).

Even though research has begun to measure research competencies, this field is still in the beginning of its development (ibid.). Therefore, in 2011, the BMBF[2] launched a large funding initiative called KoKoHs ("Kompetenzmodelle und Instrumente der Kompetenzerfassung im Hochschulsektor—Validierungen und methodische Innovationen" or "Modeling and Measuring Competencies in Higher Education").[3]

The research project LeScEd (an acronym for "**Le**arning the **Sc**ience of **Ed**ucation"), which was funded in the line of KoKoHs, comprised researchers from the University of Koblenz and Landau, the University of Göttingen, the Free University of Berlin, the University of Freiburg, and the University of Education Freiburg (cf. Schladitz et al., 2013).

The project was dedicated to examine three facets of research orientation of university students and doctoral candidates:

- Knowledge and mastery of procedures and methods of social sciences;
- Scientific argumentation and communication;
- Epistemological beliefs about the nature of knowledge and knowing.

Central research questions of this project include modeling of scientific thinking and working of university students in educational disciplines.

The chapters of this book present results from the sub-project #3 of LeScEd in which beliefs with a special focus on mathematics using the theoretical framework of epistemological beliefs were studied. The global intentions are (a) to identify epistemological beliefs about mathematics as a science; and subsequently (b) to develop the instruments to do so reliably and economically.

In a first step, we investigated the assumption that the usual categories used in traditional questionnaires (i.e., self-report Likert-scale items) have to be differentiated. We argue that epistemological judgments can be grounded in different beliefs as well as other cognitive arguments (e.g., Stahl 2011), within a considerable range of sophistication. In later steps, we developed a questionnaire with open items and used it in several large samples. This will be exemplified by the topics of *certainty of knowledge* and *justification of knowledge.*

[2]**B**undes**m**inisterium für **B**ildung und **F**orschung—Federal Ministry of Education and Research.

[3]See, for example: https://www.kompetenzen-im-hochschulsektor.de/.

### 1.3.1 Overview of All Publications from Sub-Project #3 of LeScEd

The results from sub-project #3 have been published in several peer-reviewed journal articles, book chapters, and conference proceedings. A complete list is given below, sorted by their publication medium.

**(A) Book chapters**

(A1) **Rott, B.**, Leuders, T., & Stahl, E. (2013). "Is Mathematical Knowledge Certain?—Are You Sure?" Development of an Interview Study to Investigate Epistemological Beliefs. In S. Zh. Praliev & H.-W. Huneke (Eds.), *Current Problems of Modern University Education* (pp. 167–174). Kazakh Notional Pedagogic Abai University. Ulagat.

(A2) **Rott, B.** (2017). "Is Mathematical Knowledge Certain?—Are You Sure?" A Fictitious Classroom Discussion. In M. Stein (Ed.), *A Life's Time for Mathematics Education and Problem Solving. Festschrift on the Occasion of András Ambrus' 75th Birthday* (pp. 364–369). Münster: WTM.

**(B) Peer-reviewed journal articles**

(B1) **Rott, B.**, Leuders, T., & Stahl, E. (2014). "Is Mathematical Knowledge Certain?—Are You Sure?" An Interview Study to Investigate Epistemic Beliefs. *mathematica didactica, 37*, 118–132.

(B2) **Rott, B.**, Leuders, T., & Stahl, E. (2015). Assessment of mathematical competencies and epistemic cognition of pre-service teachers. *Zeitschrift für Psychologie, 223*(1), 39–46.

(B3) **Rott, B.** & Leuders, T. (2016). Inductive and deductive justification of knowledge: Flexible judgments underneath stable beliefs in teacher education. *Mathematical Thinking and Learning, 18*(4), 271–286.

(B4) **Rott, B.** & Leuders, T. (2017). Mathematical competencies in higher education: Epistemological beliefs and critical thinking in different strands of pre-service teacher education. *Journal for Educational Research Online, 9*(2), 113–134.

**(C) Journal articles**

(C1) Schladitz, S., **Rott, B.**, Winter, A., Wischgoll, A., Groß Ophoff, J., Hosenfeld, I., Leuders, T., Nückles, M., Renkl, A., Stahl, E., Watermann, R., Wirtz, M., & Wittwer, J. (2013). LeScEd—Learning the Science of Education. Research Competence in Educational Sciences. In S. Blömeke & O. Zlatkin-Troitschanskaia (Eds.), *The German funding initiative "Modeling and Measuring Competencies in Higher Education": 23 research projects on engineering, economics and social sciences, education and generic skills of higher* *education students. (KoKoHs Working Papers, 3)* (pp. 82–84). Berlin & Mainz: Humboldt University & Johannes Gutenberg University.

(C2) **Rott, B.** (2015). Was glauben Lernende wie unser Wissen entsteht und wie erfasst man dieses Thema?—Bericht über ein Symposium zu epistemologischen Überzeugungen. *Mitteilungen der GDM, 99*, 49–51.

**(D) Peer-reviewed conference proceeding articles**

(D1) **Rott, B.**, Leuders, T., & Stahl, E. (2015). Epistemological Judgments in Mathematics: an Interview Study Regarding the Certainty of Mathematical Knowledge. In C. Bernack-Schüler, R. Erens, A. Eichler, & T. Leuders (Eds.), *Views and Beliefs in Mathematics Education: Proceedings of the MAVI 2013 Conference* (pp. 227–238). Berlin: Springer.

(D2) **Rott, B.** & Leuders, T. (2016). Mathematical critical thinking: The construction and validation of a test. In C. Csíkos, A. Rausch, & J. Szitányi (Eds.), *Proceedings of the 40th Conference of the International Group for the Psychology of Mathematics Education, Vol. 4* (pp. 139–146). Szeged, Hungary: PME. [Research Report]

**(E) Conference proceeding articles**

(E1) **Rott, B.**, Leuders, T. & Stahl, E. (2014). Zusammenhang von epistemologischen Überzeugungen und kritischem Denken bei Studierenden des Lehramts Mathematik. In GEBF (Ed.), *GEBF-Abstractband „Die Perspektiven verbinden": 2. Tagung der GEBF in Frankfurt am Main, 03.–05. März 2014* (p. 686). Goethe-Universität Frankfurt am Main.

(E2) **Rott, B.**, Leuders, T., & Stahl, E. (2014). „Wie sicher ist Mathematik?"– epistemologische Überzeugungen und Urteile und warum das nicht dasselbe ist. In J. Roth & J. Ames (Eds.), *Beiträge zum Mathematikunterricht 2014* (pp. 1011–1014). Münster: WTM.

(E3) **Rott, B.**, Leuders, T., & Stahl, E. (2014). Belief Structures on Mathematical Discovery—Flexible Judgments Underneath Stable Beliefs. In S. Oesterle, C. Nicol, P. Liljedahl, & D. Allan (Eds.), *Proceedings of the Joint Meeting of PME 38 and PME-NA 36, Vol. 6* (p. 213). Vancouver, Canada: PME. [Oral Communication]

(E4) **Rott, B.**, Leuders, T., & Ostermann, A. (2015). Facetten mathematischer Forschungskompetenz—epistemologische Überzeugungen und kritisches Denken. In GEBF (Ed.), *GEBF-Abstractband „Heterogenität.Wert.Schätzen": 3. Tagung der GEBF in Bochum, 11.–13. März 2015* (p. 26). Ruhr-Universität Bochum.

(E5) **Rott, B.** & Leuders, T. (2015). Neue Ansätze zur Erfassung epistemologischer Überzeugungen von Studierenden. In F. Caluori, H. Linneweber-Lammerskitten, & C. Streit (Eds.), *Beiträge zum Mathematikunterricht 2015* (pp. 752–755). Münster: WTM.
(E6) **Rott, B.**, Groß Ophoff, J., & Leuders, T. (2017). Erfassung der konnotativen Überzeugungen von Lehramtsstudierenden zur Mathematik als Wissenschaft und als Schulfach. In U. Kortenkamp & A. Kuzle (Eds.), *Beiträge zum Mathematikunterricht 2017* (pp. 1101–1104). Münster: WTM.
(E7) **Rott, B.** & Leuders, T. (2017). Mathematical Critical Thinking: A Question of Dimensionality. In B. Kaur, W. K. Ho, T. L. To & B. H. Choy (Eds.), *Proceedings of the 41st Conference of the International Group for the Psychology of Mathematics Education, Vol. 1* (p. 263). Singapore: PME. [Oral Communication]

In this book, the main results from sub-project #3 are summarized. The chapters of this book, which address different parts of our research project (a qualitative pre-study, a quantitative pilot study, and a quantitative main study), are each mainly based on one or two publications, see Table 1.2. As the table shows, the four peer-reviewed journal articles are especially important for the contents of this book. Therefore, their abstracts are presented below.

**Table 1.2** Chapters of the book at hand and the publications (see list above) which were used (almost) entirely in those chapters. From publications listed in parentheses, only small amounts were used or they had been extended to the respective journal articles

| Chapter | **Title** | **Publications** |
|---|---|---|
| 1 | Introduction | (C1) |
| 2 | Interviews Regarding Epistemological Beliefs | B1 (E2, A1), D1, A2 |
| 3 | Measuring Mathematical Critical Thinking | D2, E7 |
| 4 | Critical Thinking and Epistemological Beliefs of Pre-Service Teachers | B2 (E1, E5) |
| 5 | Epistemological Beliefs and Critical Thinking in Pre-Service Teacher Education | B4 (E4) |
| 6 | Inductive and Deductive Justification of Knowledge | B3 (E3) |
| 7 | Summary and Outlook | E6 |

**Abstracts of the peer-reviewed journal articles**

(B1) Rott, B., Leuders, T., & Stahl, E. (2014). "Is Mathematical Knowledge Certain?—Are You Sure?" An Interview Study to Investigate Epistemic Beliefs. *mathematica didactica, 37*, 118–132.

*Abstract*: The goal of the study presented in this article is to identify epistemic beliefs about mathematics as a scientific discipline with a focus on *certainty of knowledge*. The topic is discussed from both a psychological point of view and from the philosophy of mathematics. An interview study was conducted to investigate on the question whether domain-general conceptualizations of the certainty dimension are applicable to the domain mathematics. The recorded positions and arguments ask for a domain-specific conceptualization of certainty.

(B2) Rott, B., Leuders, T., & Stahl, E. (2015). Assessment of Mathematical Competencies and Epistemic Cognition of Pre-Service Teachers. *Zeitschrift für Psychologie, 223*(1), 39–46.

*Abstract*: Assessment in higher education requires multifaceted instruments to capture competence structures and development. The construction of a competence model of pre-service mathematics teachers' mathematical abilities and epistemic beliefs allows for comparing different groups at different stages of their studies. We investigated 1st and 4th semester students with respect to their epistemic beliefs on the certainty of mathematical knowledge (assessed by both denotative and connotative judgments) and their mathematical abilities (defined as critical thinking with respect to mathematical problem situations). We show that students' beliefs change during the first four semesters, and that the level of critical thinking does not depend on belief orientation but goes along with the level of sophistication of the epistemic judgments uttered by the students.

*Keywords*: epistemic beliefs, mathematical competencies, sophisticated beliefs, critical thinking

(B3) Rott, B. & Leuders, T. (2016). Inductive and deductive justification of knowledge: Flexible judgments underneath stable beliefs in teacher education. *Mathematical Thinking and Learning, 18*(4), 271–286.

*Abstract*: Personal epistemological beliefs are considered to play an important role for processes of learning and teaching. However, research on personal epistemology is confronted with theoretical issues as there is conflicting evidence regarding the structure, stability, and context-dependence of epistemological beliefs. We give evidence how theoretical and methodological issues can partly be resolved by distinguishing between relatively stable "epistemological beliefs" and situation-specific "epistemological judgments." A qualitative content analysis of a series of semistructured interviews (study 1) with pre-service teachers, teachers, and teacher educators as well as a statistical analysis of pre-service teachers' extensive answers in questionnaires (study 2), both on the topic of "mathematical discovery," reveal not only beliefs of the participants but also different qualities of judgments. Therefore, in further research both aspects of beliefs should be considered in a more differentiated manner when categorizing belief structures.

(B4) Rott, B. & Leuders, T. (2017). Mathematical competencies in higher education: Epistemological beliefs and critical thinking in different strands of pre-service teacher education. *Journal for Educational Research Online, 9*(2), 113–134.

*Abstract*: Assessment in higher education requires multifaceted instruments to capture competency structures and development. We investigate two aspects of competencies of pre-service mathematics teachers: a certain aspect of mathematical abilities (critical thinking with respect to mathematical problem situations) and epistemological beliefs (assessed by both belief position and belief justification). We investigated 463 students from two universities with respect to both aspects of competencies. We show that students' belief position and justification are independent and can be assessed independently. Whereas belief position is not correlated with the number of the students' semesters, their course of studies, and their mathematical abilities, belief justification is indeed correlated with these factors.

*Keywords*: Epistemological Beliefs; Critical Thinking; Mathematical Competencies

## References

Arner, D. G. (1972). *Perception, reason, and knowledge: An introduction to epistemology*. Glenview, IL: Scott, Foresman and Company.

Baxter Magolda, M. B. (1992). *Knowing and reasoning in college: Gender-related patterns in students' intellectual development*. San Francisco: Jossey Bass.

Belenky, M. F., Clinchy, B. M., Goldberger, N. R., & Tarule, J. M. (1986). *Women's ways of knowing: The development of self, voice and mind*. New York: Basic Books.

Bierman, D. B. (2008). A qualitative study of the epistemological interplay between teachers and students in a high stakes testing environment. Dissertation. Rhode Island College. Retrieved from http://digitalcommons.ric.edu/cgi/viewcontent.cgi?article=&context=etd.

Borg, S. (2010). Language teacher research engagement. *Language Teaching, 43*(4), 391–429.

Briell, J., Elen, J., Verschaffel, L., & Clarebout, G. (2011). Personal epistemology: nomenclature, conceptualizations, and measurement (chapter 2). In J. Elen, E. Stahl, R. Bromme, & G. Clarebout (Eds.), *Links between beliefs and cognitive flexibility—lessons learned* (pp. 7–36). Dordrecht: Springer.

Bromme, R., & Kienhues, D. (2008). Allgemeinbildung. In W. Schneider & M. Hasselhorn (Eds.), *Handbuch der Pädagogischen Psychologie* (pp. 619–628). Göttingen: Hogrefe.

Butchvarov, P. (1970). *The concept of knowledge.* Evanston: Northwestern University Press.

Clarebout, G., Elen, J., Luyten, L., & Bamps, H. (2001). Assessing epistemological beliefs: Schommer's questionnaire revisited. *Educational Research and Evaluation, 7*(1), 53–77.

Dionné, J. (1984). The perception of mathematics among elementary school teachers. In J. Moser (Ed.), *Proceedings of the sixth annual meeting of the PME-NA* (pp. 223–228). Madison (WI): University of Wisconsin.

Elen, J., Stahl, E., Bromme, R., & Clarebout, G. (2011). *Introduction* (chapter 1). In J. Elen, E. Stahl, R. Bromme, & G. Clarebout (Eds.), *Links between beliefs and cognitive flexibility—lessons learned* (pp. 1–6). Dordrecht: Springer.

Ernest, P. (1989). The impact of beliefs on the teaching of mathematics. In P. Ernest (Ed.), *Mathematics teaching: The state of the art* (pp. 249–254). London: Falmer Press.

Fishbein, M., & Ajzen, I. (1975). *Belief, attitude, intention, and behavior*. Reading: Addison-Wesley.

Furinghetti, F. (1998). Around the term 'belief'. In M. Hannula (Ed.), *Proceedings of MAVI-3 Workshop (Current state of research on mathematical beliefs VII)* (pp. 24–29). Helsinki: University of Helsinki. Department of Teacher Education. Research report 198.

Furinghetti, F., & Pehkonen, E. (2002). Rethinking characterisations of beliefs. In G. Leder, E. Pehkonen, & G. Törner (Eds.), *Beliefs: A hidden variable in mathematics education?* (pp. 39–57). Dordrecht: Kluwer Academic.

Goldin, G. A. (2002). Affect, meta-affect, and mathematical belief structures. In G. C. Leder, G. C., E. Pehkonen, & G. Törner (Eds.), *Mathematics education library: Vol. 31. Beliefs: A hidden variable in mathematics education?* (pp. 59–72). Dordrecht: Kluwer Academic.

Green, T. F. (1971). *The activities of teaching*. New York: McGraw-Hill.

Grigutsch, S., Raatz, U., & Törner, G. (1998). Einstellungen gegenüber Mathematik bei Mathematiklehrern. *Journal für Mathematik-Didaktik, 19*(1), 3–45.

Groß Ophoff, J., & Rott, B. (2017). Educational research literacy—special issue editorial. *Journal for Educational Research Online, 9*(2), 5–10.

Hart, L. E. (1989). Describing the affective domain: Saying what we mean. In D. B. McLeod & V. M. Adams (Eds.), *Affect and mathematical problem solving—A new perspective* (pp. 37–45). New York: Springer.

Hofer, Barbara K. (2000). Dimensionality and disciplinary differences in personal epistemology. *Contemporary Educational Psychology, 25*, 378–405.

Hofer, B. K. (2001). Personal epistemology research: Implications for learning and teaching. *Journal of Educational Psychology Review, 13*(4), 353–383.

Hofer, B. K., & Pintrich, P. R. (1997). The development of epistemological theories: Beliefs about knowledge and knowing and their relation to learning. *Review of Educational Research, 67*(1), 88–140.

Höttecke, D. (2001). Die Vorstellungen von Schülern und Schülerinnen von der „Natur der Naturwissenschaften". *Zeitschrift für Didaktik der Naturwissenschaften, 7,* 7–23.

King, P. M., & Kitchener, K. S. (1994). *Developing reflective judgment: Understanding and promoting intellectual growth and critical thinking in adolescents and adults.* San Francisco: Jossey-Bass.

Kloosterman, P., & Stage, F. K. (1992). Measuring beliefs about mathematical problem solving. *School Science and Mathematics, 91*(3), 109–115.

Kuhn, D. (1991). *The skills of argument.* Cambridge: Cambridge University Press.

KMK—Kultusministerkonferenz (2004). Standards für die Lehrerbildung: Bildungswissenschaften. Beschluss der Kultusministerkonferenz. [Standards for Teacher Education: Educational Sciences. Resolution of the Standing Conference of Education Ministers]. Retrieved from http://www.kmk.org/fileadmin/veroeffentlichungen_beschlues-se/2004/2004_12_16-Standards-Lehrerbildung.pdf.

McLeod, D. B. (1989). Beliefs, attitudes, and emotions: New views of affect in mathematics education. In D. B. McLeod & V. M. Adams (Eds.), *Affect and mathematical problem solving—A new perspective* (pp. 245–258). New York: Springer.

Muis, K. R. (2004). Personal Epistemology and mathematics: A critical review and synthesis of research. *Review of Educational Research, 74*(3), 317–377.

Murphy, P. K., & Mason, L. (2006). Changing knowledge and beliefs. In P. A. Alexander & P. H. Winne (Eds.), *Handbook of educational psychology* (2nd ed., pp. 305–324). London: Lawrence Erlbaum Publishers.

Neumann, I., & Kremer, K. (2013). Nature of Science und epistemologische Überzeugungen—Ähnlichkeiten und Unterschiede. *Zeitschrift für Didaktik der Naturwissenschaften, 19,* 209–232.

Pehkonen, E. (1999). Conceptions and images of mathematics professors on teaching mathematics in school. *International Journal of Mathematical Education in Science and Technology, 30*(3), 389–397.

Perry, W. G. (1968). *Patterns of development in thought and values of students in a liberal arts college: A validation of a scheme.* Cambridge: Harvard University.

Perry, W. G. (1970). *Forms of intellectual and ethical development in the college years: A scheme.* New York: Holt, Rinehart and Winston.

Philipp, R. A. (2007). Mathematics teachers' beliefs and affect (Chapter 7) In F. K. Lester (Ed.), *Second handbook of research on mathematics teaching and learning* (pp. 257–315). Charlotte: Information Age.

Piaget, J. (1950). *Introduction à l'epistemologie genetique.* Paris: Presses Univ. de France.

Piaget, J. (1974). *Abriß der genetischen Epistemologie.* Freiburg: Walter Verlag Olten.

Rott, B. (2018). Problem solving in the classroom: How do teachers organize lessons with the subject problem solving? In A. Ambrus & É. Vásárhelyi (Eds.), *Problem solving in mathematics education—Proceedings of the 19th ProMath conference.* Budapest: Eötvös Loránd University Press.

Rott, B. (2020). Teachers' behaviors, epistemological beliefs, and their interplay in lessons on the topic of problem solving. *International Journal of Science and Mathematics Education, 18,* 903–924.

Schladitz, S., Rott, B., Winter, A., Wischgoll, A., Groß Ophoff, J., Hosenfeld, I., Leuders, T., Nückles, M., Renkl, A., Stahl, E., Watermann, R., Wirtz, M., & Wittwer, J. (2013). LeScEd—Learning the science of education. research competence in educational sciences. In S. Blömeke & O. Zlatkin-Troitschanskaia (Eds.), *The German funding initiative "modeling and measuring competencies in higher education": 23 research projects on engineering, economics and social sciences, education and generic skills of higher education students. (KoKoHs Working Papers, 3)* (pp. 82–84). Berlin & Mainz: Humboldt University & Johannes Gutenberg University.

Schoenfeld, A. H. (1985). *Mathematical problem solving*. Orlando, FL: Academic Press.

Schoenfeld, A. H. (1992). Learning to think mathematically. In D. A. Grouws (Ed.), *Handbook for research on mathematics teaching and learning* (pp. 334–370). New York: MacMillan.

Schommer, M. (1990). Effects of beliefs about the nature of knowledge on comprehension. *Journal of Educational Psychology, 82*, 498–504.

Schommer, M. (1993). Epistemological development and academic performance among secondary students. *Journal of Educational Psychology, 85*(3), 406–411.

Schommer, M. (1998). The role of adults' beliefs about knowledge in school, work, and everyday life (Ch. 7). In M. C. Smith & T. Pourchot (Eds.), *Adult learning and development* (pp. 127–143). London: Lawrence Erlbaum.

Schraw G, Bendixen LD, Dunkle ME. (2002) Development and evaluation of the Epistemic Belief Inventory (EBI) In B. K. Hofer & P. R. Pintrich (Eds.), *Personal Epistemology: The psychology of beliefs about knowledge and knowing* (pp. 261–275). Mahwah, NJ: Lawrence Erlbaum Associates.

Schwitzgebel, E. (2015). Belief. In E. N. Zalta (Ed.), *The stanford encyclopedia of philosophy* (Summer 2015 Edition). Retrieved from https://plato.stanford.edu/archives/sum2015/entries/belief/.

Stahl, E. (2011). The generative nature of epistemological judgments: Focusing on interactions instead of elements to understand the relationship between epistemological beliefs and cognitive flexibility (Chapter 3). In J. Elen, E. Stahl, R. Bromme, & G. Clarebout (Eds.), *Links between beliefs and cognitive flexibility—Lessons learned* (pp. 37–60). Dordrecht: Springer.

Thompson, A. G. (1992). Teachers' beliefs and conceptions: A synthesis of the research. In D. A. Grouws (Ed.), *Handbook of research on mathematics teaching and learning* (pp. 127–146). Reston: National Council of Teachers of Mathematics.

Törner, G., & Pehkonen, E. (1999). Teachers' beliefs on mathematics teaching—Comparing different self-estimation methods—A case study. Retrieved from https://duepublico.uni-duisburg-essen.de/servlets/DerivateServlet/Derivate-5246/mathe91999.pdf.

Trautwein, U., & Lüdtke, O. (2007). Epistemological beliefs, school achievement, and college major: A large-scale longitudinal study on the impact of certainty beliefs. *Contemporary Educational Psychology, 32*, 348–366.

Tremp, P., & Futter, K. (2012). Forschungsorientierung in der Lehre. Curriculare Leitlinie und studentische Wahrnehmungen. [Research Orientation in Teaching. Curricular Guidelines and Students' Perception.] In T. Brinker & P. Tremp (Eds.), *Einführung in die Studiengangentwicklung* (pp. 69–79). Bielefeld: Bertelsmann.

Tsai, C.-C. (1998a). An analysis of scientific epistemological beliefs and learning orientations of Taiwanese eight graders. *Science Education, 82*(4), 473–489.

Tsai, C.-C. (1998b). An analysis of Taiwanese eighth graders' science achievement, scientific epistemological beliefs and cognitive structure outcomes after learning basic atomic theory. *International Journal of Science Education, 20*(4), 413–425.

Wegner, C., Weber, P., & Fischer, O. S. (2012). Epistemologische Überzeugungen – eine Untersuchung zur Beeinflussbarkeit der Auffassungen über die Natur des Wissens. *news&science. Begabtenförderung und Begabungsforschung, 30*(1), 49–56.

Wegner, E., & Nückles, M. (2011). Die Wirkung hochschuldidaktischer Weiterbildung auf den Umgang mit widersprüchlichen Handlungsanforderungen. *ZFHE, 6*(3), 171–188.

# Interviews Regarding Epistemological Beliefs

# 2

To identify epistemological beliefs regarding mathematics, an interview study was conducted. The interviewees were selected to represent as broad a sample as possible, ranging from university students (pre-service teachers of mathematics) to in-service teachers, working mathematicians, and full professors (of mathematics and mathematics education), see Table 2.1.

To prepare for the conduction and analyses of these interviews, the literature from mathematics and philosophy of mathematics was scanned to identify philosophical positions regarding mathematical epistemology as well as empirical studies focusing on this topic.

The first part of this chapter is based on a (peer-reviewed) journal article (Rott et al. 2014c)[1] with small additions from a (peer-reviewed) conference proceedings chapter (Rott et al. 2015a)[2] and summarizes the above-mentioned interview study in a scientific form. The second part of this chapter was previously published as a book chapter in a *Festschrift* (Rott 2017)[3] and presents results from the interview study in the form of a fictitious classroom dialogue in the tradition of Pólya and Lakatos.

---

[1]The article was published in the journal *mathematica didactica*, Franzbecker Verlag. This publisher leaves its authors with full control over their content and allows them to reuse their publications in a different form.

[2]This article first appeared in the *MAVI (Mathematical Views) 2013 Proceedings*, published by Springer; rights to reprint (parts of) the article have been obtained.

[3]The book was published in the WTM-Verlag (Wissenschaftliche Texte und Medien), which allows its authors to reuse and republish their work without any limitations.



B. Rott, *Epistemological Beliefs and Critical Thinking in Mathematics*, Freiburger Empirische Forschung in der Mathematikdidaktik, https://doi.org/10.1007/978-3-658-33539-7_2

**Table 2.1** Overview of the part of the interview (1: *certainty*, 2: *justification*, 3: *ontology*) and the publications (Rott et al., YEAR) in which the interview data was used

| I. | S. | Status | Part | Date | Publication |
|---|---|---|---|---|---|
| C. K. | m | pre-service teacher (math) | (1) | 03.12.2012 | |
| D. S. | m | pre-service teacher (math) | (1) | 03.12.2012 | |
| H. K. | m | pre-service teacher (math) | (1) | 03.12.2012 | |
| K. N. | f | pre-service teacher (math) | (1) | 03.12.2012 | |
| | | | (2) | 20.12.2012 | |
| | | | (3) | 28.05.2013 | |
| T. W. | m | pre-service teacher (math) | (1) | 03.12.2012 | (2013, 2014c, 2015a) |
| | | | (2) | 28.05.2013 | |
| | | | (3) | 28.05.2013 | |
| T. H. | m | pre-service teacher (math) | (1) | 03.12.2012 | (2014a) |
| | | | (2) | 28.05.2013 | |
| | | | (3) | 28.05.2013 | |
| A. R. | m | mathematician | (1) | 27.12.2012 | (2013, 2014a, c, 2015a) |
| | | | (2) | 27.12.2012 | |
| | | | (3) | 27.12.2012 | |
| D. B. | m | in-service teacher (math) | (1) | 29.12.2012 | |
| | | | (2) | 29.12.2012 | (2016) |
| | | | (3) | 29.12.2012 | |
| T. B. | m | professor (math educ.) | (1) | 25.01.2013 | (2014a) |
| | | | (2) | 25.01.2013 | |
| B. G. | f | pre-service teacher (math) | (1) | 29.01.2013 | (2013, 2014a, c, 2015a) |
| | | | (2) | 29.01.2013 | (2014b, 2016) |
| | | | (3) | 29.01.2013 | |
| R. E. | m | in-service teacher (math) | (1) | 14.03.2013 | |
| | | | (2) | 14.03.2013 | |
| | | | (3) | 14.03.2013 | |
| S. W. | m | professor (math) | (1) | 04.04.2013 | (2013, 2014c, 2015a) |
| | | | (2) | 04.04.2013 | (2016) |
| | | | (3) | 04.04.2013 | |
| P. S. | f | PhD student (not math) | (1) | 15.05.2013 | (2014c) |

(continued)

**Table 2.1** (continued)

| I. | S. | Status | Part | Date | Publication |
|---|---|---|---|---|---|
| | | | (2) | 15.05.2013 | |
| | | | (3) | 15.05.2013 | |
| A. H. | f | pre-service teacher (math) | (1) | 28.05.2013 | |
| | | | (2) | 28.05.2013 | (2016) |
| | | | (3) | 28.05.2013 | |
| C. P. | f | pre-service teacher (math) | (1) | 27.08.2013 | (2014c) |
| | | | (2) | 27.08.2013 | (2014b, 2016) |
| | | | (3) | 27.08.2013 | |
| K. E. | m | professor (economics) | (1) | 05.01.2014 | |
| | | | (2) | 05.01.2014 | |
| | | | (3) | 05.01.2014 | |

*Legend* initials (I), sex (S), and status of the interviewees

## 2.1 Interview Study—Certainty of Mathematical Knowledge

### 2.1.1 Introduction

To understand the conditions, the processes and the outcome of tertiary education it is necessary to capture the knowledge and beliefs of students by developing theories and instruments that describe relevant competence areas (Blömeke et al. 2013). For teacher education, several instruments for assessing competencies on a large scale have been developed in the past decade (Blömeke et al. 2013; Baumert et al. 2010; Hill et al. 2005). Despite these developments, many aspects of teachers' competencies still lead to a controversial discussion. In particular, there are many open questions as to the structure of the beliefs of future mathematics teachers (Pajares 1992; Grigutsch et al. 1998; Bernack-Schüler et al. in preparation).

The principal goals of the project reported in this chapter (which is embedded within the research initiative "Modeling and Measuring Competencies in Higher Education", cf. Blömeke et al., 2013) are to identify epistemological beliefs about mathematics as a scientific discipline; and subsequently to develop the instruments to do so reliably and economically. In this chapter, we focus on a first step of this process: A preliminary interview study was conducted to identify epistemological beliefs of people dealing with mathematics (e.g., teacher students or mathematicians) in detail. Even though there exists a multitude of questionnaires and other instruments to measure epistemological beliefs, we chose interviews for reasons of validity. Our research will be exemplified by the topic of certainty of knowledge which is central to mathematics (Kline 1980).

A person's beliefs are his/her "[p]sychologically held understandings, premises, or propositions about the world that are thought to be true." (Philipp 2007, p. 259) They filter his/her perceptions and direct his/her actions (cf. Philipp 2007). For example, beliefs about mathematics influence the person's mathematical problem solving performance (e.g., Schoenfeld 1992) and his/her acquisition of mathematical knowledge (see Muis 2004, p. 339 ff. for an overview of related studies).

## 2.1.2 The Structure of Epistemological Beliefs

Epistemology is a branch of philosophy that deals with the nature of human knowledge, including its limits, justifications, and sources (cf. Arner 1972, ch. I). Research on personal epistemology as a branch of both psychology and education research examines the development of epistemological beliefs (which are beliefs about the nature of knowledge and knowing) and the way these beliefs filter perceptions, influence learning processes, and direct actions.

In mathematics education, research on beliefs rarely uses the terms "epistemological" or "epistemic" (e.g., these are only briefly mentioned by Thompson 1992, or Philipp 2007). Instead, research on this topic is sometimes assessed under the construct of "mathematical world views" (cf. Schoenfeld 1992; Grigutsch et al. 1998).

Early studies in the 1970s and 80s from developmental psychology modeled personal epistemology as a one-dimensional sequence of stages in which "individuals move through some specified sequence in their ideas about knowledge and knowing, as their ability to make meaning evolves." (Hofer 2001, p. 356) Since the 1990s, especially researchers from educational psychology consider personal epistemology as a system of more or less independent epistemological beliefs which allow for a more differentiated way of describing personal epistemology development as well as for a discipline-specific understanding of epistemology (cf. ibid., p. 361). A widely accepted structure for such a system of beliefs was proposed by Hofer and Pintrich (1997). According to Hofer and Pintrich, there

**Table 2.2** Areas and dimensions of epistemological beliefs (Hofer and Pintrich 1997)

| *Nature of knowledge* (what one believes knowledge is) | | *Nature or process of knowing* (how one comes to know) | |
|---|---|---|---|
| *Certainty of knowledge* | *Simplicity of Knowledge* | *Source of knowledge* | *Justification of knowledge* |
| Assumptions about the certainty or tentativeness of knowledge | Assumptions whether knowledge consists of simple facts vs. it is a complex network of information | Assumptions whether knowledge originates outside the self and resides in external authority vs. assumptions of the self as knower, with the ability to construct knowledge | Assumptions about how to evaluate knowledge claims, the role of empirical evidence, experts, and how to justify knowledge claims |

are two general areas of epistemological beliefs (*nature of knowledge* and *nature or process of knowing*) with two dimensions each (see Table 2.2).

A growing amount of psychological research presents relationships between epistemological beliefs and various aspects of learning. Examples (see Table 2.3 for these studies' methodological approaches) include the way in which college students' epistemological beliefs influence their processing of information and their monitoring of comprehension (e.g., Mason and Boscolo 2004), their academic performance (e.g., Schommer et al. 1997), conceptual change (e.g., Qian and Alvermann 1995), cognitive processes during learning (e.g., Kardash and Howell 2000), learning processes within computer-based scenarios and with the Internet (e.g., Hofer 2004a), and their engagement in learning (e.g., Hofer and Pintrich 1997).

Furthermore, there is evidence that students' beliefs about knowledge and academic concepts depend on teaching style and epistemological beliefs of their teachers (e.g., Buelens et al. 2002; Hofer 2004b; Tsai 1998). It is generally assumed that more sophisticated epistemological beliefs are related to more adequate learning strategies and therefore better learning outcomes (cf. Hofer and Pintrich 1997; Stahl 2011).

Trying to summarize recent research on epistemological cognition, one could say that most studies use self-report instruments like questionnaires to investigate correlations of epistemological beliefs with various aspects of learning and reception of information.

**Table 2.3** An overview of the cited studies with a focus on methodology

| Authors | Methodology |
|---|---|
| Buelens (2002) | They used a thirty-item questionnaire to study the correspondence of conceptions of knowledge and learning with conceptions of instruction |
| Hofer (2004a) | She used think-aloud protocols and retrospective interviewing to examine how students engage in epistemological metacognitive processes during online searching for simulated class assignments |
| Hofer (2004b) | She used field notes from observations and interviewed students at the beginning and the end of this observation that lasted one semester |
| Hofer and Pintrich (1997) | They conducted a meta study and summarized the results of other studie. |
| Kardash and Howell (2000) | They used a 42-item version of Schommer's (1990) "Epistemological Belief Questionnaire" to investigate the effects of epistemological beliefs and topic-specific beliefs on undergraduates' cognitive and strategic processing of a dual-positional text (thinking aloud) |
| Mason and Boscolo (2004) | They used an instrument by Kuhn et al. (2000) to measure the "epistemological understanding" by judging situations. The total scores could range from 15 (absolutist positions in all judgment domains) to 45 (evaluativist positions in all judgment domains) |
| Qian and Alvermann (1995) | They used the "Epistemological Belief Questionnaire" (Schommer, 1990) to correlate epistemological beliefs and learned helplessness with conceptual understanding and application reasoning in conceptual change learning |
| Schommer et al. (1997) | They used Schommer's (1990) questionnaire |
| Tsai (1998) | He used a questionnaire to identify 20 "information-rich" students that were then interviewed |

However, basic research on epistemological beliefs is still needed, as there exist conflicting data that cannot be explained with traditional theories about epistemological beliefs (cf. Bromme et al. 2008). Even though most researchers have conceived this construct as general and rather stable, growing empirical evidence showed that epistemological beliefs are less coherent, more discipline-related and more context-dependent than it was hitherto assumed (cf. Hofer 2000).

For example, Muis et al. (2011) analyzed the epistemological beliefs of students enrolled in a statistics course. They showed that "slight changes in context influence what epistemological beliefs are activated, which can subsequently influence learning." (ibid., p. 516)

Stahl (2011) claims that it is necessary to distinguish between relatively stable *epistemological beliefs* and situation specific *epistemological judgments* when examining this construct in more detail. *Epistemological judgments* are defined

> [...] as learners' judgments of knowledge claims in relation to their beliefs about the nature of knowledge and knowing. They are generated in dependency of specific scientific information that is judged within a specific learning context. [...] [A]n epistemological judgment might be a result of the activation of different cognitive elements (like epistemological beliefs, prior knowledge within the discipline, methodological knowledge, and ontological assumptions) that are combined by a learner to make the judgment. (Stahl 2011, p. 38 f.)

Stahl (2011) elaborates these theoretical considerations of a generative nature of epistemological judgments with fictitious examples. Three persons with different backgrounds (content knowledge, methodological knowledge, ontological assumptions, epistemological beliefs, etc.) in physics each judge the claim that the distance between sun and earth is 149.60 million kilometers. In this chapter we intend to support this assumption by empirical examples.

In mathematics education the terms "personal epistemology" and "epistemological beliefs" are rarely used. Instead, research on this topic is assessed under the construct of beliefs (cf. Muis 2004, p. 322). Muis summarizes several studies dealing with beliefs about mathematics:

> The majority of research that has examined students' beliefs about mathematics suggests that students at all levels hold nonavailing[4] beliefs. In general, when asked about the certainty of mathematical knowledge, students believe that knowledge is unchanging. The use and existence of mathematics proofs support this notion, and students believe the goal in mathematics problem solving is to find the right answer. [...] (Muis 2004, p. 330)

Researchers investigating beliefs about mathematics as a discipline deal with opposing perceptions of mathematics: process-orientation versus rule-orientation, dynamic versus static interpretation, formal versus informal discipline, or its applicability (cf. Muis 2004; Grigutsch et al. 1998).

---

[4]To avoid a negative connotation, Muis (2004, p. 323 f.) does not use the common labels "naïve—sophisticated" or "inappropriate—appropriate" from psychological and educational research. Instead she suggests to use the labels "nonavailing—availing" for beliefs that are associated with better learning outcomes ("availing"), and for beliefs that have no influence or a negative influence on learning outcomes ("nonavailing").

The *global intentions* of our research project are (a) to identify epistemological beliefs about mathematics as a science and (b) to develop the instruments to do so economically (cf. Muis 2004, p. 354). The *research intentions* for this chapter are (i) to identify epistemological beliefs about mathematics as a science (especially regarding "certainty of knowledge"), and (ii) to provide empirical evidence that supports the theoretical differentiation between epistemological beliefs and epistemological judgments.

Therefore, we take a close look to the certainty of mathematical knowledge.

### 2.1.3 Certainty of Knowledge

In this paragraph, we present the conceptualization of epistemological beliefs regarding the *certainty of knowledge* in different models of epistemic cognition as well as related research results. We further discuss the certainty of knowledge in a domain-specific view regarding mathematics drawing on the philosophy of mathematics.

#### 2.1.3.1 Certainty of Knowledge in Models of Epistemic Cognition

*Certainty of knowledge* is a (sub-)dimension that is included in almost every conception of personal epistemology (cf. Hofer and Pintrich 1997). In all of these models, a strong belief in truth and certainty is a sign of a naïve standpoint. For example, in Perry's (1970) original model of epistemic development, a *dualistic view* that sees knowledge claims either "right" or "wrong" is the least sophisticated stage. In higher stages of development, knowledge is considered to be constructed by humans and to be uncertain. In Schommer's (1990) questionnaire, strong beliefs in certainty of knowledge are considered as a sign of naivety. Hofer and Pintrich describe the dimension *certainty of knowledge* as follows:

> The degree to which one sees knowledge as fixed or more fluid appears throughout the research, again with developmentalists likely to see this as a continuum that changes over time, moving from a fixed to a more fluid view. At lower levels, absolute truth exists with certainty. At higher levels, knowledge is tentative and evolving. [...] (Hofer and Pintrich 1997, p. 119 f.)

Hofer and Pintrich (1997, p. 98) are consistent with Perry's ideas and other models of epistemic cognition in their description of this dimension: Absolutistic manifestations of this dimension go along with the view of "knowledge as certain and [the belief] that authorities have all the answers." As they become more sophisticated, learners recognize that knowledge may be uncertain and that authorities may possibly not know the truth. At even higher levels of sophistication, knowledge is regarded as "uncertain and contextual, but it is now possible to coordinate knowing and justification to draw conclusions across perspectives." (ibid, p. 101) Expert authority is critically evaluated at this stage.

Researchers slowly realize that epistemological beliefs are not general across all domains (cf. Hofer 2006). However, the degree to which this conceptualization of certainty is applicable to domains like mathematics is yet to be evaluated and will be discussed in this paper.

The following is a summary of examples for research results using the conceptualization of certainty as illustrated above. In her first questionnaire study with 86 junior college students, Schommer (1990, p. 502 f.) identified a certainty dimension. "The more the students believed in certain knowledge, the more likely they were to write [inappropriately] absolute conclusions [in comprehension tasks]." In a study with 326 first-year college students, Hofer (2000, p. 402) identified a significant relation between the certainty dimension and academic achievement. "As expected, epistemological beliefs, at least on the dimension of certainty/simplicity of knowledge, are correlated with academic performance. This was the case whether the dimension was discipline specific or general." Trautwein and Lüdtke (2007) conducted a study with 1094 students who were tested twice—in their final year of upper secondary school and two years later, after their first year at university. They showed that their certainty scale—even after controlling for other variables—was a significant predictor of the final school grade. Certainty beliefs also predicted the specific fields of study at university.

#### 2.1.3.2 Certainty of Mathematical Knowledge

The *certainty of mathematical knowledge* differs from the certainty of knowledge in other disciplines and is therefore a deeply discussed topic in the philosophy of mathematics. There are good arguments for both—the certainty and the uncertainty of mathematical knowledge.

On the one hand, especially from a historical perspective, mathematical knowledge is regarded as certain, because of formal proofs and deductive reasoning with respect to strict rules and axioms (cf. Heintz 2000, p. 52 ff.; Hoffmann 2011, p. 1 ff.). This is a characteristic, which is mostly unique for mathematics; compared to other (natural) sciences, mathematical knowledge does not depend on observations and experiments but only on logical conclusions. Arner (1972, p. 116) points out that "[t]he historical importance of mathematics as a paradigm of a priori truth needs no emphasis." According to Arner, with the discovery of non-Euclidean geometries, all references to data and applications were abandoned.

> Next it was demonstrated that not only geometry but other branches as well can be developed by the deductive method, from a relatively few assumptions, and likewise without reliance upon empirical data. As a result all pure mathematics is found to be abstract, in the sense of being independent of any particular application. [...] It was here [in 'Principia Mathematica' by Whitehead and Russell] proved that the initial assumptions of mathematics can all be dispensed with, except the definitions. The truths of mathematics follow merely from definitions which exhibit the meaning of its concepts, by purely logical deduction. Judgment of such mathematical truth is, thus, completely and exclusively analytic; no synthetic judgment, a priori or otherwise, is requisite to knowledge of pure mathematics. The content of the subject consists entirely of the rigorous logical analysis of abstract concepts, in entire independence of all data of sense or modes of intuition. (Arner 1972, p. 117 f.)

On the other hand, the assumption of absolute certainty revealed several flaws in the more recent history of mathematics:

*Theoretical perspective*: (1a) In formal proofs, each conclusion can be drawn by rules of inference relying on preceding sentences; this way, every mathematical statement can be traced back to very elementary rules and axioms. However, it is impossible to justify these axioms and the discovery of non-Euclidean geometries has shown that different stipulations of axioms can lead to divergent mathematics. Therefore, different prerequisites can lead to different outcomes concerning the same mathematical object. (1b) The finding of contradictory derivations from axioms (Russell's paradox, 1901) led to the attempt of establishing formal rules of derivation and to the effort of finding a complete and consistent set of axioms by Russell himself ("Principia Mathematica"), Hilbert and others. However, this attempt was regarded as a failing because of Gödel's incompleteness theorems in 1931, which stated that any sufficiently strong formal system cannot be both consistent and complete (cf. Hoffmann 2011, p. 52 ff.)

*Ontological perspective*: (2) Mathematical knowledge cannot be objectively justified. There is no proof of a Platonian world to which mathematical results can be referred to.

*Empirical perspective*: (3a) Mathematical knowledge is spread by publishing in journals and to ensure the correctness of submitted proofs, mathematicians review each other's submissions. However, this review process cannot guarantee the identification of all inaccuracies or even incorrect parts that might be hidden in these proofs. Often, mathematical work is so specialized that only a handful of experts is able to comprehend it and the history of mathematics is full of examples of accepted proofs that were discovered to be wrong years after their publication (e.g., the proof of the Four Color Theorem, cf. Wilson 2002). (3b) Finally, a growing number of mathematical results are achieved with the help of computers (e.g., the famous proof of the Four Color Theorem) and no mathematician is able to verify them without trusting the machines as well as hoping for error-free hard- and software (cf. Borwein and Devlin 2011, p. 8 ff.).

The discussion about the fundamentals of mathematics during the last century has led to a panoply of different philosophical stances that coexist in the community of mathematicians and even within single persons, since there is no way of proving the validity of one position or another (e.g., Barrow 1992; Bedürftig and Murawski 2012). As a foundation for our research, this wealth of arguments is the basis for creating rich interview situations.

### 2.1.4 Methodology

In this paragraph, we explain our choice to conduct interviews and we describe the development and implementation of our interview setting. To assess epistemological beliefs about mathematics in depth, we chose different key questions from the philosophy of mathematics, which provided us with a rich background of subject-specific theoretical positions and arguments. As an example of our research and as a central aspect of epistemological beliefs, we looked into the key question of *certainty of knowledge*.

#### 2.1.4.1 Capturing Epistemological Beliefs: Methodological Considerations

A common method of measuring epistemological beliefs is through the use of questionnaires. Very influential ones are the "Epistemological Questionnaire" (EQ) by Schommer (1990) and a questionnaire to measure teachers "views of mathematics and its structure" by Grigutsch et al. (1998). Both are widely used among researchers and have inspired numerous variations and follow-up versions. However, questionnaires raise methodological issues regarding their validity and their effectiveness (cf. Stahl 2011; Muis 2004) as well as psychometric problems that all self-report instruments suffer from (Greene and Yu 2014). Stahl summarizes methodological issues:

> [A]ll attempts to date to develop a questionnaire with strong reliability and validity have brought little success. The main problem is seen in the unstable factor structure of the instruments. Another problematic aspect concerns the items which are often indirectly related to epistemological beliefs. (Stahl 2011, p. 41 f.)

Muis (2004) identifies additional difficulties with questionnaires in their effectiveness and in their capability of measuring general as well as domain-specific epistemological beliefs. She concludes that questionnaires should be developed that "address epistemological beliefs effectively" and that there should be "two types of questionnaires [...] [o]ne should be specific to each domain of interest, and one should be general across domains." (ibid., p. 354)

Therefore, both Stahl as well as Greene and Yu suggest the use of interviews, observations, or other instruments rather than classical questionnaires to understand epistemological beliefs of learners in detail. Following these recommendations in the study presented here, we use interviews to measure domain-specific (mathematics) epistemological beliefs.

#### 2.1.4.2 Development and Implementation of the Interviews

To initiate the interviews, we presented quotes of representatives of opposing epistemological positions to the interviewees and asked them to relate themselves to these statements. This proved to generate more insight into our subjects' beliefs than direct questioning (e.g., "Do you think that mathematical knowledge is certain?"), because this way, our interviewees had different positions to refer to and got an impression of the breadth of the argument. After this initial prompt, we asked further questions and intervened with information contrary to the subjects' positions to further identify their lines of reasoning (see below). Presenting specific contexts and thus specifying the type of knowledge that is discussed is vital to guarantee measurement validity (cf. Greene and Yu 2014, p. 23).

During the first phase of data collection, we optimized our selection of quotes to start the interviews with as well as the pursuing questions. For example, we added explicit headlines to our quotes ("mathematical knowledge is certain/uncertain", see Table 2.4) instead of solely presenting the quotes to emphasize the two positions. Starting with only one intervention for all students ("is mathematical knowledge really certain?"), we also developed additional interventions for the various positions and reasons our interviewees raised. The data collection ended when the authors felt that the data were saturated. From a methodological point of view, the development of the research design with respect to successively analyzed data and the collection of data until saturation is reached, is similar to the approach of Grounded Theory (cf. Strauss and Corbin 1996).

So far, the first author interviewed 9 pre-service teachers of mathematics (students at the University of Education, Freiburg), 2 in-service teachers of mathematics, 1 professional mathematician, 1 professor of mathematics education, 1 professors of mathematics, as well as 1 professor of economics, and 1 Ph.D. student from veterinary medicine (see Table 2.1). The interview data has

**Table 2.4** Starting positions for "Certainty of Mathematical Knowledge"

| Mathematical Knowledge is Certain | Mathematical Knowledge is Uncertain |
|---|---|
| "In mathematics, knowledge is valid forever. A theorem is never incorrect. In contrast to all other sciences, knowledge is accumulated in mathematics. […]<br>It is impossible, that a theorem that was proven correctly will be wrong from a future point of view. Each theorem is for eternity."<br>(Albrecht Beutelspacher) [2001, p. 235; translated by the first author] | "The issue is […] whether mathematicians can always be absolutely confident of the truth of certain complex mathematical results […].<br>With regard to some very complex issues, truth in mathematics is that for which the vast majority of the community believes it has compelling arguments. And such truth may be fallible.<br>Serious mistakes are relatively rare, of course."<br>(Alan H. Schoenfeld) [1994, p. 58 f.] |

been analyzed and validated consensually within intense discussions among the researchers.

In our sample setting, the interview started like this: "These are two positions of mathematicians regarding the certainty of mathematical knowledge. With which position can you identify yourself? Please give reasons for your answer." (See Table 2.4 for the two positions.) Further questions were: "Can you explain your position on the basis of your own mathematical experience?" and "Please compare the certainty of mathematical knowledge to that of other sciences, for example to physical, linguistic, or educational knowledge." These questions proved to be helpful in revealing our interviewees' content knowledge, methodological knowledge, and ontological assumptions.

If an interviewee settled on "mathematical knowledge is *certain*", we confronted him/her with information about the scientific review process and the history of the Four Color Theorem (cf. Wilson 2002): In 1879, Kempe submitted a proof for this problem that was accepted by the community of mathematicians, only to be shown to be false by Heawood eleven years later. The currently valid proof by Apple and Haken is so complicated and so large (more than 400 pages) that no one can be sure of its correctness.

If a subject expressed that "mathematical knowledge is *uncertain*", we gave him/her additional information about deductive reasoning and presented an easy proof of the Pythagorean Theorem. "How can a theorem like this one, with hundreds of proofs, countless validations and practical applications be regarded as uncertain?"

Overall, applying all the questions and prompts described here as well as questions that arose from the situation, the interviews lasted between 5 and 20 min.

### 2.1.5 Results

After interviewing university students of mathematics as well as professional mathematicians and professors of mathematics, we can present the following stances regarding the certainty of mathematical knowledge and underpin them with supporting statements.[5]

We found several representatives of either the position that "mathematical knowledge is certain" or that "mathematical knowledge is uncertain" as well as some interviewees that were undecided. For both "certain" and "uncertain" we found participants that showed substantially different argumentations due to their backgrounds. In the following, we present three examples for each position showing graduated levels of sophistication.

#### 2.1.5.1 Interviewees Judging that "Mathematical Knowledge is Certain"

*T.W. is a pre-service mathematics teacher* in his third semester (age 23). For him, mathematical knowledge is certain, "the first quote of Beutelspacher is more likely correct in my view." He thinks of proofs as inevitable and irrefutable and adds: "How can there possibly be errors in mathematics?" Confronted with the historical episode of the four color theorem, T.W. admits "Of course, there can be errors, [...] but it got proven eventually, didn't it?" He mentioned the Pythagorean Theorem as an example of an everlasting theorem and did not know about the mathematical review process (and its limitations) before our intervention.

*P.S. is a doctoral student in veterinary medicine* in her second year (age 28). She shows a very sophisticated knowledge of the way scientific results are gained and justified in the natural sciences like biology and medicine. But she has no concrete idea of the way mathematical results are obtained. Asked to compare the certainty of mathematical knowledge to that of other scientific disciplines, P.S. answers: "I'd say it is more certain than biological knowledge. Very similar to that of physical knowledge, because both, mathematics and physics, [...] build on ideas and logical thinking but less on tangible things." Compared to medical

[5] The interviews have all been conducted in German; quotes have been translated by the author.

research, she imagines mathematical results to be not as easily refuted because experiments cannot prove them wrong.

*A.R. is a mathematician* who just obtained his diploma degree (age 31). He was able to activate more content knowledge than the previous two cited interviewees. For example, he referred to Andrew Wiles' proof of Fermat's Last Theorem as well as the Riemann Conjecture. His claim of certainty was based on the deductive way of reasoning in mathematics. Even though he knew about possible flaws in the review process, A.R. did attribute this error not to mathematics: "Humans are fallible. [...] There might be errors in proofs which are accepted by many people. [...] But when a theorem is proved correctly from the axioms by formal rules of derivation then it will last for eternity."

*Interpretation:* The three interviewees above all represent the position that mathematical knowledge is certain, but the degree of sophistication varies significantly. T.W. has naïve beliefs of certainty; for him, there is not a real possibility for doubt in mathematical knowledge. P.S. is an expert in the process of knowledge generation in the natural sciences, but she has no knowledge about mathematical research. She cannot argue with possible flaws of mathematic-specific results as she can do with results from biological research. A.R. does not only know about the scientific review process and its flaws, but also about the specific way of reasoning in mathematics and about examples from the history of mathematics. Of those three, A.R. represents the most sophisticated epistemological beliefs regarding mathematical knowledge.

#### 2.1.5.2 Interviewees Judging that "Mathematical Knowledge is Uncertain"

*B.G. is a pre-service mathematics teacher* who just finished her university studies (age 25). She thinks that not only mathematical but all knowledge is uncertain, because "for me, there is always the possibility that someone figures out that something is not quite correct. A theorem might be proven and checked but there is always the possibility of finding an aspect that it may not be correct." In comparison to other scientific disciplines, she thinks that mathematical knowledge is quite certain; but general uncertainty remains: "I generally do not agree to statements referring to 'ever' or 'never'."

*C.P. is a pre-service mathematics teacher* in her fourth semester (age 23). She thinks that mathematical knowledge is uncertain, "because it has occurred several times in the history of mathematics that a theorem or its proof has proven to be false." After further questions, she admits that all mathematical results taught at school are beyond doubt but there can always be errors in complex results at universities and in mathematical research.

*S.W. is a professor of mathematics* (age 41). He used arguments referring to the review process as well as ontological positions: "Mathematical knowledge cannot be definitely certain because that would imply an infallible system of rules with an otherworldly justification. Mathematics would need a justification outside of the human sphere and outside of the mathematical discourse, a realm that could be observed and described. That there is such a realm, such a sphere, I am very skeptical about."

*Interpretation:* These three interviewees all claimed that mathematical knowledge is uncertain. Whereas B.G. could only rely on fundamental conceptions ("nothing is certain"), C.P. was able to support her position with knowledge about the review process and events from the history of mathematics. S.W. is the most sophisticated representative of these three. He does not only use arguments regarding the review process and historical events, but he also refers to the ontology of mathematical knowledge.

#### 2.1.5.3 Questionnaire Data

Previously to the interviews, the participants were asked to fill in a questionnaire (CAEB, Stahl and Bromme 2007). One of the items from this questionnaire is presented in Fig. 2.1. The interviewees' responses were all consistent with their statements in the interviews.

In combination with the results from the interviews, this questionnaire data supports the theoretical claim of Stahl (2011):

> In a questionnaire with rating scales, [these] persons would give the same answer. However, the conclusion that their responses are an expression for comparable epistemological beliefs would be wrong. Their epistemological judgments are built on different cognitive elements to evaluate the knowledge claim. (Stahl 2011, p. 49)

| **Mathematics as a scientific discipline from my point of view is …** | | | | | | | | | |
|---|---|---|---|---|---|---|---|---|---|
| T.W. | Certain | ☐ | ☑ | ☐ | ☐ | ☐ | ☐ | ☐ | Uncertain |
| P.S. | Certain | ☐ | ☑ | ☐ | ☐ | ☐ | ☐ | ☐ | Uncertain |
| A.R. | Certain | ☑ | ☐ | ☐ | ☐ | ☐ | ☐ | ☐ | Uncertain |
| B.G. | Certain | ☐ | ☐ | ☐ | ☐ | ☐ | ☐ | ☑ | Uncertain |
| C.P. | Certain | ☐ | ☐ | ☐ | ☐ | ☐ | ☑ | ☐ | Uncertain |
| S.W. | Certain | ☐ | ☐ | ☐ | ☐ | ☑ | ☐ | ☐ | Uncertain |

**Fig. 2.1** Interviewees' reactions to the questionnaire, item 22 of 24: certainty

### 2.1.6 Discussion and Outlook

*Certainty of Knowledge* is an important category in nearly all models of epistemic cognition. It is generally assumed that beliefs of certainty are a sign for absolutistic or unreflected beliefs whereas beliefs of uncertainty hint at unsophisticated beliefs about the nature of knowledge and knowing. However, there are serious doubts regarding the domain-generality of such assumptions (cf. Hofer 2006).

Drawing on domain-specific beliefs concerning the nature of mathematical knowledge, we conducted an interview study with students as well as professors of mathematics. The analyses of the interviews reveal a breadth of arguments with respect to the epistemological question whether mathematical knowledge is certain or uncertain. We encountered unreflected as well as reflected representatives of both statements. This phenomenon contrasts the assumption from research on personal epistemology that "absolute truth exists with certainty" is valid only for "lower levels" of sophistication (Hofer and Pintrich 1997, p. 119 f.). As can be seen by our interview data (especially by the example of A.R.), this belief in the certainty of mathematical knowledge can be held in a reflected way. This is in line with recent findings of Greene and Yu (2014, p. 18) who concluded: "[...] believing 'knowledge,' as a general concept, is 'certain' is not a reliable indicator of naivety." On the other hand, there are interviewees that believe in the uncertainty of mathematical knowledge but do not use sophisticated arguments for their position. Again, referring to Greene and Yu (ibid., p. 25), "merely disagreeing with a naïve statement is not a reliable indicator of expertise".

The evaluations of the interviews also show the gain of the theoretical introduction of *epistemological judgments* (Stahl 2011). Persons that hold the same position regarding the certainty of mathematical knowledge can do so with differing backgrounds. A traditional beliefs questionnaire would not be able to detect or explain those differences. Therefore, it seems doubtful to rely on instruments that measure epistemological beliefs as a locus on a scale. It should be necessary to take into account different strands of argumentation and different backgrounds. The concept of "epistemological judgment" can be a promising starting point for developing instruments that can capture such important differences.

Building on the analyses of the interview data reported in this article, we developed an instrument with the aim to identify epistemological beliefs about mathematics effectively and reliably. Because of the known problems with common, paper-based questionnaires (see Sect. 2.1.4.1, on "Methodological Issues"), we settled for a web-based questionnaire that is able to adapt to the participants' responses. If, for example, a participant answers "mathematical knowledge is certain", he will get the "Four Color Theorem" intervention but not the "Pythagorean Theorem" intervention (see "Development and Implementation of the Interviews"). A first study using this questionnaire with about 150 pre-service teachers of mathematics is reported in Chap. 4 (see also Rott et al. 2015b).

It may also be asked whether the phenomena reported with respect to the certainty or uncertainty of mathematical knowledge also occur in other topics regarding the epistemic or ontological quality of mathematics. Indeed similar findings can be reported regarding the topic of the *justification of mathematical knowledge,* which is discussed in Chap. 6 (see also Rott et al. Rott and Leuders 2016).

## 2.2 "Is Mathematical Knowledge Certain?—Are You Sure?"

### 2.2.1 Introduction

In the interview study regarding epistemological beliefs described above (see also Rott et al 2013, 2014c), we identified different positions regarding mathematical epistemology; one of the topics that were discussed was the certainty of mathematical knowledge. To ensure a wide variety of positions and arguments, people with varying levels of mathematical education were interviewed: university students (pre-service teachers of mathematics), professionals (in-service teachers and mathematicians), as well as university professors (of mathematics and of mathematics education). In the discussion above, however, we did not explicate all belief positions and arguments from the interviews but only the extreme positions. In this sub-chapter, a closer look into the different ideas by the interviewees will be provided.

To honor András Ambrus and other Hungarian researchers, I will follow the examples of George (György) Pólya (1945) and especially Imre Lakatos (1976) and present the data in the form of a *fictitious classroom dialogue.* This way, I will summarize similar ideas held by different interviewees and assume a development from naïve to sophisticated within one dialogue—a process that would most likely take several years for most individuals.

In the following conversation, the interviewees will be pseudonymized by their initials (Table 2.1). All answers from the "students" presented below are English translations of direct quotes from the interviews that have been arranged in a way to simulate a group discussion. The interventions by the "teacher" are ideas that have been presented by the interviewer (the author) to stimulate the participants to show the arguments behind their positions; those statements are not direct quotes but have been adapted to lead to a more coherent conversation.

### 2.2.2 The Classroom Discussion

TEACHER: Let us talk about the certainty of mathematical knowledge. Do you believe that the theorems from your mathematics schoolbooks are true? All of them, without any doubt? What about mathematics that is done at universities? To start our discussion, I present two contradictory quotes from mathematicians regarding this topic. With which position can you identify yourself? Please, give reasons for your answer.

> In mathematics, knowledge is valid forever. A theorem is never incorrect. In contrast to all other sciences, knowledge is accumulated in mathematics. [...] It is impossible, that a theorem that was proven correctly will be wrong from a future point of view. Each theorem is for eternity. (Beutelspacher 2001, p. 235; translated by BR)

> The issue is [...] whether mathematicians can always be absolutely confident of the truth of certain complex mathematical results [...]. With regard to some very complex issues, truth in mathematics is that for which the vast majority of the community believes it has compelling arguments. And such truth may be fallible. Serious mistakes are relatively rare, of course. (Schoenfeld 1994, p. 58 f.)

T.H.: Mhm, the first one, Beutelspacher, is right. Generally, majorities are right. However, in mathematics, theorems are discovered by individuals. When a theorem exists, it is true. That is just the way it is.

B.G.: I cannot agree to [the position of] Beutelspacher, therefore, I am for Schoenfeld. That a theorem should be true for eternity, I cannot agree to that. For me, there is always the possibility to find out that something is not right or that it is not entirely true.

TEACHER: Okay, I think that we have to move on from simply believing one position or the other. Instead, we have to use arguments. That even mathematicians can err sometimes, is important to point out. Consider the following example of "Fermat numbers" from the history of mathematics.

> In 1640, in a letter to Marin Mersenne, Pierre de Fermat expressed the conjecture that all numbers in the following sequence are prime numbers: 3, 5, 17, 257, 65537, ... To obtain these numbers, he used this formula: $F_n = 2\hat{}(2^n) + 1$, resulting in $F_0 = 2+1 = 3$, $F_1 = 4+1 = 5$, ... For the first five numbers, Fermat's conjecture was verified. However, these numbers get so large quickly, that it was nearly impossible to check

the conjecture properly. Almost 100 years later, in 1732, Leonard Euler discovered that the next number in that sequence, $F_5 = 2^{32} + 1$ was divisible by 641. Actually, with the help of modern computers, no other prime number with $n>4$ has been found in this sequence until today.

T.W.: Of course, there can be mistakes. But if someone finds a proof, than [the theorem] is irrefutably true. Therefore, the quote by Beutelspacher is correct for mathematics.

TEACHER: Speaking of proofs, consider the Pythagorean Theorem. It is known for more than 2000 years, and there are hundreds of different proofs as well as countless examples that confirm its truth. Does this settle our argument?

B.G.: I have good faith in this theorem and I use it myself. However, I still believe that it could happen that someone finds something that contradicts it. Maybe not in the [Euclidean] plane, but if you go to other planes.

TEACHER: Okay, I agree that the age of a theorem or its proof is not really meaningful. Look at the early years of the four color theorem:

In 1852, Francis Guthrie came up with the four color problem. He assumed that for any map with continuous regions that should be colored so that no two adjacent regions have the same color, four colors suffice. However, he was not able to find a proof for his hypothesis and asked Augustus De Morgan to help him prove it. De Morgan was also unable to find a proof, but posed the problem in a magazine. The problem drew some attention and several mathematicians tried to solve it. For example, in 1879 and 1880, Alfred Kempe and Peter Guthrie Tait each published proofs that were widely acclaimed. In 1890 and in 1891, however, those proofs were shown to be incorrect by Percy Heawood and Julius Petersen, respectively. Both false proofs stood unchallenged for 11 years each.

A.R.: Okay, [but I still] identify myself with Beutelspacher. Humans are fallible. Because of that, there can be errors in proofs and those proofs can also be accepted by several people with only later recognizing that it was wrong. But I start with the premise that if a theorem has been proven correctly, starting with the axioms and using the mathematical rules of reasoning; than the theorem holds true for eternity. Look at Fermat's Last Theorem and its proof by Andrew Wiles. The theorem has been proven, then errors have been found and it took several years to fix those errors. Nowadays, the proof is accepted as correct.

TEACHER: That are two arguments that we should keep apart: Mathematical axiomatics and the review process.

T.B.: The proof of Andrew Wiles is a very nice example. […] This is, as far as I know, a proof of more than 400 pages. And this proof builds on—I don't know how many—thousands of pages of [mathematical] statements. Here, it starts to become uncertain because humans are fallible and these things cannot be checked with 100% certainty.

TEACHER: Is this the "majority of the community" that Schoenfeld refers to? The social component of mathematics? Think of, for example, the method of proving by contradiction which has not been accepted by some mathematicians, most notably Luitzen Brouwer. Or let us take another look at the four color theorem.

In 1976, the four color theorem was proved by Kenneth Appel and Wolfgang Haken. They narrowed down the number of maps that had to be checked if four colors suffice to almost 2000 types of maps. To check all those maps by hand would take a mathematicians lifetime, not to mention the time of reviewers. Therefore, a computer was used to check all those cases. It was the first major theorem that was proved with the assistance of computers. This caused an intense discussion whether the hard- and software of computers was trustworthy.

T.B.: Okay, regarding this, the community is involved and there is a social component to mathematics like Schoenfeld claimed.

A.R.: [I agree that] this proof [by Wiles] is very complex and only a handful of mathematicians understands it completely. This is at the current edge of mathematical knowledge. However, in the coming decades, this proof will be deeply analyzed. And the new language that Wiles has invented will become familiar to mathematicians. And then, the proof will be accepted for all eternity.

S.W.: It is a good question, if mathematics can be done error-free. With regard to the review process, I am not so skeptical. I believe that the community [of mathematicians] is really thorough and at least the not too complex things have been thought through often, so that there is a certain degree of consistency in the [mathematical] system of reasoning. And, compared to other sciences, in some sense, mathematics is more pure. Therefore, mathematical knowledge is more certain than knowledge of other sciences. This does not mean, however, that mathematics leads to absolute truth.

A.R.: In scientific disciplines like physics, you have hypotheses that need to be tested in laboratories. Physical knowledge is falsifiable; you can only show that it is not wrong—under certain conditions. In mathematics, it is not necessary to repeat tests and to use experimental setups. In mathematics, we know the basic elements and can build from thereon. In physics, you have a complex system and you try to understand the rules of this system without knowing all the factors. This is the other way round, you cannot build your system logically, but there is a system—the nature—and we try to recognize the basic elements.

S.W.: But this does not mean that this system of rules that is used [in mathematics] is 100% correct and infallible. Additionally, basically, you could speak of an infallible system of rules only then, if there would be an otherworldly or transcendent justification. A justification outside of the human sphere, outside of the mathematical discourse. This would mean, and I think that Beutelspacher indicates this implicitly, that there would exist something like a mathematical sphere, a Platonic sphere, that we could look at and describe. And I am very skeptical regarding the existence of such a sphere.

T.B.: There are certain areas of mathematics in which we can speak of absolute certainty. Namely the mathematics that is related to reality. In my view, this includes rational numbers and after that theory begins. And for mathematical objects like 0.999... there are different theories that lead to different conclusions and statements. For example, 0.999... is equal to 1 in our standard analysis [that is taught in schools] but smaller than 1 in non-standard analysis that includes infinitesimals.

TEACHER: Okay, let us summarize our discussion. It seems that neither statement—mathematical knowledge is certain vs. uncertain—should be believed unthinkingly. There are compelling arguments for both positions. Mathematical knowledge is certain—and especially more certain than knowledge from other scientific disciplines—because of a unique reasoning process and proofs that can be traced back to the most basic elements, the axioms. Additionally, a rigorous review process ensures that the possible errors in this process of reasoning are minimized. However, mathematical knowledge—at least advanced and complex theorems—can also be regarded as uncertain, because mathematician can and do make errors in proving and reviewing. Also, there are limitations to mathematical reasoning because there is no justification outside the mathematical discourse, no Platonic sphere. An additional argument that we have not spoken of would be Gödel's incompleteness theorems that demonstrate the inherent limitations of formal axiomatic systems.

### 2.2.3 Conclusion

*Certainty of Knowledge* is an important category in nearly all models of personal epistemology. It is generally assumed that beliefs of certainty are a sign for naïve, absolutistic, or unreflected beliefs whereas beliefs of uncertainty hint at sophisticated beliefs about the nature of knowledge and knowing (cf. Hofer and Pintrich 1997). One of the major findings of our interview study was that this basic assumption is not true—at least for the domain of mathematics: The belief position regarding certainty of knowledge (mathematical knowledge is certain vs. uncertain) is not related to the according argumentation (whether the belief position is backed-up by naïve/inflexible or by reflected/sophisticated arguments). There are, for example, students that think mathematical knowledge is certain but do not have convincing arguments to back-up their position. On the other hand, there are students that use very sophisticated arguments to justify their belief that mathematical knowledge is certain.

## References


Arner, D. G. (1972). *Perception, reason, and knowledge—An introduction to epistemology.* London: Scott, Foresman and Company.

Baumert, J., Kunter, M., Blum, W., Brunner, M., Voss, T., Jordan, A., et al. (2010). Teachers' mathematical knowledge, cognitive activation in the classroom, and student progress. *American Educational Research Journal, 47,* 133–180.

Barrow, J. D. (1992). *Pi in the sky.* London: Oxford University Press.

Bedürftig, T., & Murawski, R. (2012). *Philosophie der Mathematik.* De Gruyter (2nd edn.).

Bernack-Schüler, C., Leuders, T., Holzäpfel, L., & Renkl, A. (in preparation). *Getting the right picture of mathematics by doing mathematics—Changing beliefs through individual mathematical inquiry.*

Beutelspacher, A. (2001). *Pasta all'infinito – Meine italienische Reise in die Mathematik.* München: dtv.

Blömeke, S., Zlatkin-Troitschanskaia, O., Kuhn, C., & Fege, J. (Eds.). (2013a). *Modeling and measuring competencies in higher education.* Rotterdam, the Netherlands: Sense Publishers.

Blömeke, S., Suhl, U., & Döhrmann, M. (2013b). Assessing strengths and weaknesses of teacher knowledge in Asia, Eastern Europe and Western countries: Differential item functioning in TEDS-M. *International Journal of Science and Mathematics Education, 11*(4), 795–817.

Borwein, J., & Devlin, K. (2011). *Experimentelle Mathematik – Eine beispielorientierte Einführung.* Heidelberg: Spektrum.

Bromme, R., Kienhues, D., & Stahl, E. (2008). Knowledge and epistemological beliefs: An intimate but complicate relationship. In M. S. Khine (Ed.), *Knowing, knowledge and beliefs. Epistemological studies across diverse cultures* (pp. 423–441). New York: Springer.

Buelens, H., Clement, M., & Clarebout, G. (2002). University assistants' conceptions of knowledge, learning and instruction. *Research in Education, 67,* 44–57.

Greene, J. A., & Yu, S. B. (2014). Modeling and measuring epistemic cognition: A qualitative re-investigation. *Contemporary Educational Psychology, 39,* 12–28.

Grigutsch, S., Raatz, U., & Törner, G. (1998). Einstellungen gegenüber Mathematik bei Mathematiklehrern. *Journal für Mathematik-Didaktik, 19*(1), 3–45.

Heintz, B. (2000). *Die Innenwelt der Mathematik.* [The Inner World of Mathematics] Wien: Springer.

Hill, H. C., Rowan, B., & Loewenberg Ball, D. (2005). Effects of teachers' mathematical knowledge for teaching on student achievement. *American Educational Research Journal, 42*(2), 371–406.

Hofer, B. K. (2000). Dimensionality and disciplinary differences in personal epistemology. *Contemporary Educational Psychology, 25,* 378–405.

Hofer, B. (2001). Personal epistemology research: Implications for learning and teaching. *Journal of Educational Psychology Review, 13*(4), 353–383.

Hofer, B. K. (2004a). Epistemological understanding as a metacognitive process: Thinking aloud during online searching. *Educational Psychologist, 39*(1), 43–55.

Hofer, B. K. (2004b). Exploring the dimensions of personal epistemology in differing classroom contexts: Students interpretations during the first year of college. *Contemporary Educational Psychology, 29,* 129–163.

Hofer, B. K. (2006). Beliefs about knowledge and knowing: Domain specificity and generality. *Educational Psychology Review, 18,* 67–76.

Hofer, B. K., & Pintrich, P. R. (1997). the development of epistemological theories: Beliefs about knowledge and knowing and their relation to learning. *Review of Educational Research, 67*(1), 88–140.

Hoffmann, D. (2011). *Grenzen der Mathematik.* Heidelberg: Spektrum.

Kardash, C. M., & Howell, K. L. (2000). Effects of epistemological beliefs and topic-specific beliefs on undergraduates' cognitive and strategic processing of dual-positional text. *Journal of Educational Psychology, 92*(3), 524–535.

Kline, M. (1980). *Mathematics: The loss of certainty.* Oxford: Oxford University Press.

Kuhn, D., Cheney, R., & Weinstock, M. (2000). The development of epistemological understanding. *Cognitive Development, 15,* 309–328.

Lakatos, I. (1976). *Proofs and refutations.* Cambridge: University Press.

Mason, L., & Boscolo, P. (2004). Role of epistemological understanding and interest in interpreting a controversy and in topic-specific belief change. *Contemporary Educational Psychology, 29,* 103–128.

Muis, K. R. (2004). Personal epistemology and mathematics: A critical review and synthesis of research. *Review of Educational Research, 74*(3), 317–377.

Muis, K. R., Franco, G. M., & Gierus, B. (2011). Examining epistemic beliefs across conceptual and procedural knowledge in statistics. *ZDM Mathematics Education, 43,* 507–519.

Pajares, F. M. (1992). Teachers' beliefs and education research: Cleaning up a messy construct. *Review of Educational Research, 62*(3), 307–332.

Perry, W. G. (1970). *Forms of intellectual and ethical development in the college years: A scheme.* New York: Holt, Rinehart & Winston.

Philipp, R. A. (2007). Mathematics teachers' beliefs and affect. In F. K. Lester (Ed.), *Second handbook of research on mathematics teaching and learning* (pp. 257–315). Charlotte, NC: Information Age.

Pólya, G. (1945). *How to solve it.* Princeton: University Press.

Qian, G., & Alvermann, D. (1995). Role of epistemological beliefs and learned helplessness in secondary school students' learning science concepts from text. *Journal of Educational Psychology, 87*(2), 282–292.

Rott, B., Leuders, T., & Stahl, E. (2013). "Is mathematical knowledge certain?—Are you sure?" Development of an interview study to investigate epistemological beliefs. In S.

Zh. Praliev & H.-W. Huneke (Eds.), *Current problems of modern university education* (pp. 167–174). Kasachische Nationale Pädagogische Abai-Universität. Ulagat-Verlag.

Rott, B., Leuders, T., & Stahl, E. (2014a). „Wie sicher ist Mathematik?“ – epistemologische Überzeugungen und Urteile und warum das nicht dasselbe ist. In J. Roth & J. Ames (Eds.), *Beiträge zum Mathematikunterricht 2014* (pp. 1011–1014). Münster: WTM-Verlag.

Rott, B., Leuders, T., & Stahl, E. (2014b). Belief structures on mathematical discovery—Flexible judgments underneath stable beliefs. In S. Oesterle, C. Nicol, P. Liljedahl, & D. Allan (Eds.), *Proceedings of the joint meeting of PME 38 and PME-NA 36, Vol. 6* (p. 213). Vancouver: PME.

Rott, B., Leuders, T., & Stahl, E. (2014c). "Is mathematical knowledge certain?—Are you sure?" An interview study to investigate epistemic beliefs. *mathematica didactica, 37*, 118–132.

Rott, B., Leuders, T., & Stahl, E. (2015a). Epistemological judgments in mathematics: An interview study regarding the certainty of mathematical knowledge. In C. Bernack-Schüler, R. Erens, A. Eichler, & T. Leuders (Eds.), *Views and beliefs in mathematics education: Proceedings of the MAVI 2013 Conference* (pp. 227–238). Berlin: Springer Spektrum.

Rott, B., Leuders, T., & Stahl, E. (2015b). Assessment of mathematical competencies and epistemic cognition of pre-service teachers. *Zeitschrift für Psychologie, 223*(1), 39–46.

Rott, B., & Leuders, T. (2016). Inductive and deductive justification of knowledge: Flexible judgments underneath stable beliefs in teacher education. *Mathematical Thinking and Learning, 18*(4), 271–286.

Rott, B. (2017). "Is mathematical knowledge certain?—Are you sure?" A fictitious classroom discussion. In M. Stein (Ed.), *A life's time for mathematics education and problem solving. Festschrift on the occasion of András Ambrus' 75th Birthday* (pp. 364–369). Münster: WTM.

Schoenfeld, A. H. (1992). Learning to think mathematically: Problem solving, metacognition, and sense-making in mathematics. In D. A. Grouws (Ed.), *Handbook for research on mathematics teaching and learning* (pp. 334–370). New York: MacMillan.

Schoenfeld, A. H. (1994). Reflections on doing and teaching mathematics. In A. H. Schoenfeld (Ed.), *Mathematical thinking and problem solving* (pp. 53–69). Hillsdale, NJ: Lawrence Erlbaum Associates.

Schommer, M. (1990). Effects of beliefs about the nature of knowledge comprehension. *Journal of Educational Psychology, 82*(3), 498–504.

Schommer, M., Calvert, C., Gariglietti, G., & Bajaj, A. (1997). The development of epistemological beliefs among secondary students: A longitudinal study. *Journal of Educational Psychology, 89*(1), 37–40.

Stahl, E., & Bromme, R. (2007). The CAEB: An instrument for measuring connotative aspects of epistemological beliefs. *Learning and Instruction, 17*, 773–785.

Stahl, E. (2011). The generative nature of epistemological judgments: Focusing on interactions instead of elements to understand the relationship between epistemological beliefs and cognitive flexibility. In J. Elen, E. Stahl, R. Bromme, & G. Clarebout (Eds.), *Links between beliefs and cognitive flexibility—Lessons learned* (pp. 37–60). Dordrecht: Springer.

Strauss, A. L., & Corbin, J. M. (1996). *Grounded theory: Grundlagen qualitativer Sozialforschung*. Weinheim: Beltz.

Thompson, A. G. (1992). Teachers' beliefs and conceptions: A synthesis of the research. In D. A. Grouws (Ed.), *Handbook of research on mathematic learning and teaching* (pp. 127–146). New York: MacMillan.
Trautwein, U., & Lüdtke, O. (2007). Epistemological beliefs, school achievement, and college major: A large-scale longitudinal study on the impact of certainty beliefs. *Contemporary Educational Psychology, 32*(3), 348–366.
Tsai, C.-C. (1998). An analysis of scientific epistemological beliefs and learning orientations of Taiwanese eighth graders. *Science Education, 82*(4), 473–489.
Wilson, R. (2002). *Four colors suffice.* Princeton: Princeton University Press.

# Measuring Mathematical Critical Thinking

# 3

To accompany the identification of epistemological beliefs in the study at hand, an aspect of mathematical knowledge and competence was chosen to be assessed as well. Instead of measuring mathematical achievement in terms of students' knowledge of the content of university courses, we chose to address *critical thinking* (CT). This choice was made to capture the students' actual *use* of mathematical knowledge for a theoretically more coherent picture of their competence (cf. Weinert 2001).

This chapter is based on two (peer-reviewed) conference proceedings (Rott and Leuders 2016, 2017; the latter one had to be shortened before publication, the long version is used for this chapter)[1] and incorporates the results of five students' theses (Blum 2017; Elezkurtaj 2017; Montazeri 2017; Piecyk 2017; Raßmann 2015) that were supervised by the author (BR).

## 3.1 Mathematical Critical Thinking: Construction and Validation of a Test

### 3.1.1 Motivation

In educational psychology, CT is framed "as a set of generic thinking and reasoning skills, including a disposition for using them, as well as a commitment to using the outcomes of CT as a basis for decision-making and problem solving." (Jablonka 2014, p. 121). In his Delphi Report, Facione (1990, p. 3) understands

[1] Both articles were published in PME (International Group for the Psychology of Mathematics Education) conferences proceedings, where the copyright remains with the authors.



B. Rott, *Epistemological Beliefs and Critical Thinking in Mathematics*, Freiburger Empirische Forschung in der Mathematikdidaktik,
https://doi.org/10.1007/978-3-658-33539-7_3

CT "to be purposeful, self-regulatory judgment which results in interpretation, analysis, evaluation, and inference, as well as explanation of the evidential, conceptual, methodological, criteriological, or contextual considerations upon which that judgment is based. [...]". Though there are many different conceptualizations of CT (in philosophy, psychology, and education) the following abilities are commonly agreed upon (Lai 2011, p. 9 f.): analyzing arguments, claims, or evidence; making inferences using inductive or deductive reasoning; judging or evaluation and making decisions; or solving problems.

These characterizations imply that one cannot expect CT to be a simple trait but a complex bundle of traits that require different operationalizations with respect to different aspects and domains. Therefore, CT skills cannot be located within mathematics alone, as Facione (1990, p. 14) emphasizes: Narrowing the range of CT to a single domain would misapprehend its nature and diminish its value. Learning CT can clearly be distinguished from learning domain-specific content. However, there can be domain-specific manifestations of CT and subject contexts play an important role in learning CT (ibid.).

CT is often used to describe central educational goals—especially in higher education—and, therefore, CT skills are widely accepted as a very important part of student learning in schools as well as in universities (Lai 2011; Jablonka 2014). CT has long been supported by educators—and especially mathematics educators—, even though explicit reference to CT is rare in curricula around the world (Jablonka 2014, p. 122).

Jablonka (2014, p. 121) stresses the importance of mathematics education for the development of CT skills: "The role assigned to CT in mathematics education includes CT as a by-product of mathematics learning, as an explicit goal of mathematics education, as a condition for mathematical problem solving, as well as critical engagement with issues of social, political, and environmental relevance by means of mathematical modeling and statistics."

Because of these relationships between CT and mathematics education, further research is needed that highlights mathematics-specific approaches on CT. However, existing tests that measure CT skills are mostly domain-general and do not consider mathematics-specific features.

### 3.1.2 Measuring Critical Thinking

For example, the Ennis-Weir test of CT uses the context of general argumentation. The participants are presented with a letter that contains complex arguments. They are supposed to write a response to the given letter, defending their judgments with reasons in nine paragraphs. Each paragraph is rated with a score between –1 and 3 on the basis of a coding manual (Ennis and Weir 1985).

Another approach at measuring a disposition for CT in educational psychology is using questionnaires in which test persons indicate their consent to statements like "I would prefer complex to simple problems" or "I find satisfaction in deliberating hard and for long hours" (Cacioppo et al. 1996).

CT ability is often measured by tests like the *Watson-Glaser Critical Thinking Appraisal* (Pearson Education 2012). The Watson-Glaser CTA consists of five subtests which "require different, though interdependent, applications of analytical reasoning in a verbal context" (ibid., p. 3). In one of the subtests, for example, the probability of truth of inferences based on given information has to be rated on a 5 point Likert-scale ranging from "true" to "false".

### 3.1.3 Rationale

Tests like the ones previously mentioned are used for identifying CT in general. Recent theories of CT, however, assume that in addition to a general disposition of being a critical thinker, there are domain-specific manifestations of CT (Facione 1990; Jablonka 2014). The **research intention** described in this section is, therefore, to construct and validate a test to measure certain aspects of mathematics-specific CT. The test should be applicable for upper secondary and university students. We report on four pilot studies (three quantitative and one qualitative) to document the development of such a test.

### 3.1.4 Theoretical Background

In an attempt to measure mathematics-specific components of CT, one cannot include all aspects mentioned in the previous paragraph. Therefore, we focus on a rather basic and implicit dimension of CT that addresses the process of judgment during mathematical problem solving. This can be connected to a cognitive model by adapting and extending *dual process theory* (e.g., Kahneman 2003), which is an influential theory of human cognition (for a summary of the development of and research regarding this theory covering more than four decades, see Kahneman 2011). Doing this, Stanovich and Stanovich (2010) propose a tripartite model of thinking in which they locate CT. Similar to dual process theory, they distinguish subconscious ("type 1" or "autonomous thinking") from conscious thinking

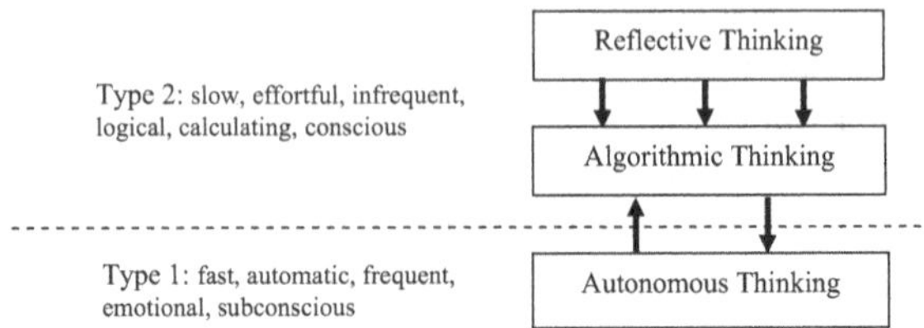


**Fig. 3.1** The tripartite model of thinking adapted from Stanovich and Stanovich (2010, p. 210); the broken horizontal line represents the key distinction in dual process theory

("type 2"). Subconscious thinking is characterized as fast, automatic, and emotional, whereas conscious thinking, which can override subconscious thinking, is characterized as slow, effortful, logical, and calculating. In addition to dual process theory, the tripartite model further differentiates conscious thinking into "algorithmic" and "reflective thinking" (see Fig. 3.1).

For Stanovich and Stanovich (2010, p. 204), this differentiation is necessary as "all hypothetical thinking involves type 2 processing [...] but not all type 2 processing involves hypothetical thinking." The authors illustrate their idea of CT with problems like item 13 (the item numbers in this chapter match the numbers in our CT test):

ITEM 13: Each of the boxes below represents a card lying on a table. Each one of the cards has a letter on one side and a number on the other side. Here is a rule: If a card has a vowel on its letter side, then it has an even number on its number side. As you can see, two of the cards are letter-side up, and two of the cards are number-side up. Your task is to decide which card or cards must be turned over in order to find out whether the rule is true or false.

Indicate which cards must be turned over.

The most common answer to task 1 is to pick A and 8, whereas A and 5 would have been the correct answer. To answer correctly, type 2 processes are necessary and the problem solvers have to consider what they can learn about the cards by picking two.

Another example for a problem that depends on the problem solver's willingness to use type 2 processes (overwriting type 1 thoughts) and to reflect upon his solution is item 7:

ITEM 7: A bat and a ball cost $ 1.10 in total. The bat costs $ 1 more than the ball. How much does the ball cost?

The spontaneous, autonomously produced answer that most problem solvers come up with is $ 0.10. A critical thinker would question this answer and realize that the ball should cost $ 0.05, whereas people who do not use CT do not evaluate their first thought and do not adapt their spontaneous solution.

Therefore, when solving mathematical problems, CT can be attributed to those processes that consciously regulate autonomous and algorithmic use of mathematical procedures. Consequently, tasks to measure mathematics-specific CT that reflect this definition should (i) reflect discipline-specific solution processes but should *not* require higher level mathematics, (ii) require a reflective component of reasoning and judgment when solving a task or evaluating the solution, and (iii) reflect an appropriate variation of difficulty within the population.

### 3.1.5 Constructing a Test for Mathematical Critical Thinking

Using the tripartite model of thinking (Fig. 3.1), CT can be operationalized by situations that demand a critical override of autonomous and algorithmic solutions by reflective and evaluative processes. Such situations can be initiated by tasks as stated above. Additionally, the tasks need to be situated in mathematics to measure domain-specific CT but should be solvable with basic level mathematics.

To compile a CT test, we collected tasks from the according literature—including item 13 (cards with vowels and even numbers) and item 7 (bat-and-ball with varied numbers in the first version: € 10.20 for both with the bat costing € 10 more than the ball). We also adapted tasks from other contexts and constructed tasks by ourselves. This way, we collected more than 30 tasks to measure mathematical CT.

In this section, we present another three examples from our list of tasks:

ITEM 9: A sequence of 6 squares made of matches consists of 19 matches (see the figure). How many matches does a sequence of 30 squares consist of?

Uncritical thinkers might infer from 6 squares to 30, resulting in $19 \cdot 5 = 95$ matches. This solution uses algorithmic thinking without the realization that the correct solution is only 91 matches because of twice counted matches after six squares each.

> ITEM 2: If the sum of the digits of an integer is divisible by three, then it cannot be a prime number.
>
> This statement is ☐ correct / ☐ incorrect

Uncritical thinkers might answer "correct" because of the "divisible by three"-rule without realizing that the prime number 3 also has a sum of digits divisible by three.

> ITEM 8: Write an equation using the variables S and P to represent the following statement: "There are six times as many students as professors at this university." Use S for the number of students and P for the number of professors.

This task by Kaput and Clement (1979) is famous for its difficulty with most persons wrongly answering "P = 6 S", revealing missing reflection.

The list of CT tasks was then used to construct a test of mathematical CT. The first version of this test did not include all tasks from our list but only 22 CT tasks. This was done to keep the time required to carry out the test below 30 min.

### 3.1.6 Validating the Test for Mathematical Critical Thinking

To control whether our test is suited to measure CT, we designed and carried out three quantitative and one qualitative pilot studies. In all studies, the tasks were rated dichotomously, 1 point for a correct answer and 0 points for a wrong answer.

#### 3.1.6.1 Pilot Study 1: CT versus Non-CT Items; Task Formulations

The first pilot study was designed to test the tasks and their formulations. It was also used to explore whether our test actually addresses CT. To investigate on the latter question, we constructed non-CT tasks that can be solved using algorithmic thinking without the need for reflection. We matched those tasks to the CT tasks with a similar context and similar computational difficulty. For example, we used the following task as a non-CT version of the bat-and-ball and matches tasks, respectively:

> ITEM 7b: You buy eight items for € 14.32 altogether. You pay with a 20 Euro note. How much change do you get?

> ITEM 9b: How many matches does the figure consist of? [The according picture shows 30 squares of matches similar to task 3, arranged in a 5 × 6 pattern.]

In total, the test of study 1 consisted of 22 CT tasks and 10 non-CT tasks. It was carried out with $n = 15$ upper secondary students (grade 11) within 40 min in October 2013.

The students correctly solved 80% of the non-CT tasks but only 58% of the CT tasks. Therefore, we concluded that our collection of tasks was suited to measure CT. As a result of this study, we removed 8 tasks from our collection due to floor or ceiling effects. Additionally, we improved the wordings of some tasks on which the students orally reported difficulties in understanding the formulations after completing the test.

#### 3.1.6.2 Pilot Study 2: Fatigue and Learning Effects Within the CT Test

The second pilot study was designed to test for decreasing concentration and possible learning effects within the 30 min of testing. We used the improved test with 14 CT tasks in two versions. Version A had the tasks 1–14 whereas version B had a different order of tasks with tasks 8–14 in the first and tasks 1–7 in the second half of the test.

In April 2014, this study was carried out with $n = 121$ pre-service teachers—students at the University of Education, Freiburg—that attended a lecture on arithmetic. The students were split into four practice groups, with two groups getting version A and the other two groups version B of the test. The results are summarized in Table 3.1. There were no statistical differences between the two groups (multiple t-tests with Bonferroni correction), indicating no fatigue or learning effects within working on the CT test.

**Table 3.1** Results of pilot study 2, mean values (standard deviations), minimum/maximum

| | Task 1–7 | | Task 8–14 | | Total | |
|---|---|---|---|---|---|---|
| | M (SD) | min/max | M (SD) | min/max | M (SD) | min/max |
| A ($n = 66$) | 3.50 (1.26) | 1/6 | 2.73 (1.32) | 1/6 | 6.23 (2.02) | 2/12 |
| B ($n = 55$) | 3.53 (1.32) | 0/7 | 2.60 (1.34) | 0/6 | 6.13 (2.19) | 2/11 |
| Total ($n = 121$) | 3.51 (1.28) | 0/7 | 2.67 (1.33) | 0/6 | 6.18 (2.09) | 2/12 |

#### 3.1.6.3 Pilot Study 3: Differentiation Between Groups

The third pilot study was conducted to examine whether the CT test is able to discriminate between different groups of students, which were either enrolled to become mathematics teachers for upper secondary schools or to become computer scientists. Some of the students were in the so-called "basic study" (semesters 1–4) whereas others were in their main study period (semesters 5 or higher). We used a shortened version (15 min) of the test from study 2 with 11 CT-items. This study was carried out with $n = 94$ students at the University of Duisburg-Essen in August 2014 (Raßmann 2015).

Our hypothesis was that students with more university experience (i.e. a higher number of semesters) would score better than students with less university experience. Table 3.2 (left side) shows the results of the students in their basic study and main study period, respectively. A t-test (after testing for normal distribution) confirmed the expected differences in favor of the more experienced students ($p_{1\text{-sided}} = 0.008 < 0.01$).

For the comparison of the students of both study programs, we did not have an assumption which program would prepare its students better for mathematical CT. The results, however, show a clear advantage for the pre-service mathematics teachers (see Table 3.2, right side, $p_{2\text{-sided}} = 0.038 < 0.05$).

**Table 3.2** Results of pilot study 3, mean values (and standard deviations)

| University experience | Item 1–11 | Study program | Item 1–11 |
|---|---|---|---|
| semester $\leq 4$ ($n = 28$) | 4.29 (2.19) | mathematics ($n = 46$) | 5.63 (2.50) |
| semester $\geq 5$ ($n = 66$) | 5.50 (2.34) | informatics ($n = 48$) | 4.67 (2.14) |
| total ($n = 94$) | 5.14 (2.31) | total ($n = 94$) | 5.14 (2.31) |

**Table 3.3** Solution rates of the five selected tasks in all three pilot studies

| Study | S1 | | S2 | | | | S3 | | |
|---|---|---|---|---|---|---|---|---|---|
| item/$n$ = | 15 | 66 | 55 | 121 | 28 | 66 | 46 | 48 | 94 |
| 13: cards | 0.20 | 0.20 | 0.13 | 0.17 | 0.14 | 0.20 | 0.28 | 0.08 | 0.18 |
| 7: bat-&-ball[a] | 0.67 | 0.45 | 0.49 | 0.47 | 0.50 | 0.62 | 0.59 | 0.58 | 0.59 |
| 9: matches | 0.53 | 0.59 | 0.58 | 0.59 | 0.61 | 0.61 | 0.74 | 0.48 | 0.61 |
| 2: divisible by 3 | 0.40 | 0.41 | 0.36 | 0.39 | 0.54 | 0.65 | 0.67 | 0.56 | 0.62 |
| 8: stud. & prof. | 0.20 | 0.06 | 0.09 | 0.07 | 0.14 | 0.24 | 0.28 | 0.15 | 0.21 |

[a]Using other numbers (€ 10.20 for both bat and ball) leads to more computational solutions and, thus, higher solution rates in study 1. We therefore used the original version (with € 1.10) in later studies

#### 3.1.6.4 Overview of Pilot Studies 1–3

Interestingly, the tasks showed very similar solution rates within all pilot studies despite the considerably different study participants. Table 3.3 presents these rates for the five tasks selected for this paper for all (sub-) populations (see above).

#### 3.1.6.5 Pilot Study 4: Task-Based Interviews

The fourth pilot study was designed to better understand the way students worked on the CT tasks. Therefore, task-based interviews with $n = 5$ pre-service mathematics teachers were conducted: three interviews at the University of Education Freiburg and two at the University of Duisburg-Essen in the period from January 2014 to September 2014. These interviews covered all 14 tasks that were used in pilot study 2. Due to space reasons, we can only present a small excerpt of these interviews.

The interviews regarding item 13 (vowels and numbers on cards) revealed that this task rather tested for knowledge (rules of mathematical reasoning) instead of CT. However, one interviewee (that previously did not have the required knowledge) solved this task correctly by reflecting on his choice of cards, showing the importance of CT for item 13.

Working on item 7 (bat-and-ball), four interviewees spontaneously said 10 Cent. However, two of them corrected their solution to 5 Cent shortly afterwards. Both told the interviewer that they found the correct solution because of checking their result. Therefore, this task is suited to reveal CT. The fifth interviewee did not express a spontaneous solution but used an equation from the beginning. It should be added that both students who checked their solution admitted that this checking was mostly due to the interview situation. This information could lead to further studies revealing situations that trigger the use of CT within students (see future prospects, below).

For item 9 (matches), the interviews showed that the wrong approach (multiplying by 5) seems to be an obvious idea. Three interviewees expressed this idea with two of them correcting their approach after a check. The other two solved this task correctly from the start. Thus, this task is also suited to test for reflective thinking.

Item 2 was solved correctly by all five interviewees with all of them showing signs of CT by expressing thoughts like: "The statement is correct. Wait, does this rule include the number 3 itself? Then it is not correct."

Item 8 was solved correctly by only one interviewee who knew the task beforehand.

In total, the task-based interviews helped us to reveal tasks that did not require CT and to confirm the use of critical or reflective thinking (in contrast to automatic or algorithmic thinking) with other tasks.

#### 3.1.6.6 Validation Study: Prompting

In 2017, the CT test was administered with two classes of students of grade 9 ($n = 59$), each of which was divided into two groups. One group each was working on the test without further information, while the other group received prompts that encourage them to "think critically" and "check [their] results" (Montazeri 2017). We expected the prompted groups to score significantly better, confirming the sensitivity of the test for reflective thinking.

Surprisingly, the test did not show significant differences between the prompted and unprompted groups, suggesting that a disposition for critical thinking cannot be activated that easily. The groups, however, were quite small so that a replication of this study seems to be appropriate.

### 3.1.7 Conclusion and Discussion

Based on the results from the quantitative pilot studies (floor and ceiling effects) and the insight provided by the interviews, we eliminated tasks (e.g., task 5). The final test consists of 14 CT tasks (including tasks 1–4) with an average time requirement of 20 min in total.

With our approach, we do not intend to include the broad range of aspects and dimensions that are currently discussed under the umbrella term "critical thinking". We also cannot contribute to the societal, curricular, and philosophical aspects of the topic. However, when one realizes the few efforts to measure CT with respect to mathematics, one could take the studies presented as an approach to pinpoint interindividual differences within CT quantitatively. Furthermore, the connection to *dual process* theory allows for a theoretical interpretation of cognitive processes that contribute to CT. In this context, our instrument can be useful in further studies to elucidate the relations of CT with other aspects such as knowledge, dispositions, and epistemological beliefs (see Rott et al. 2015). Some further theoretical connections to other relevant theories of mathematical thinking, especially to problem solving and the role of metacognition, which could not be addressed in this article, should be explored more deeply on a theoretical and empirical basis.

### 3.1.8 Additional Remarks and Future Prospects

In two studies with $n = 215$ and $n = 463$ mathematics pre-service teachers, a short version of the CT test (11 tasks) was used to measure the students' CT as a part of their mathematical knowledge base. With the test, we were able to differentiate between students of different study programs (students for upper secondary schools scored better than for primary and lower secondary schools) and different university experience (students with a higher number of semesters scored better). There were also highly significant correlations between good test scores in the CT test and the ability to justify epistemological beliefs sophisticatedly (for details, see Rott et al. 2015, or Chap. 4 in this book, respectively).

Even though this test has been validated and used, it still needs to be further explored. A next step could be a replication of the study with two groups—one group without further information, and the other group with prompts like "check your results". We are curious whether the second group will score significantly better.

Additionally, the correlation of our mathematics-specific CT test with general CT tests (e.g., by Enis and Weir 1985) should be explored as well as its correlation to students' performance in mathematical problem solving and/or argumentation. Besides improving the test, such studies could also help us to better understand the general conception of CT and its connections to problem solving, argumentation, and meta-cognition as "such association[s] remain under-theorized" (Jablonka 2014, p. 121).

## 3.2 Mathematical Critical Thinking: A Question of Dimensionality

### 3.2.1 Motivation and Additional Theoretical Background

According to *dual process theory*, cognitive activities can be distinguished into a fast, automatic, emotional, subconscious ("type 1") and a slow, effortful, logical, conscious ("type 2") subset of minds. The above mentioned characteristics of CT clearly correspond to conscious type 2 thinking which can override subconscious type 1 thinking.

Based on such a dual process model (Fig. 3.1), we propose an operationalization that enables us to reflect this structure empirically by constructing a paper-and-pencil test to measure CT (Rott and Leuders 2016). In this section, we are going to discuss test results from more than 600 students obtained in two studies to investigate on the nature of CT with respect to such cognitive processes.

As stated above, Stanovich and Stanovich (2010) proposed an extension of dual process theory, the tripartite model that further differentiates conscious thinking into "algorithmic" and "reflective thinking" (see Fig. 3.1). For Stanovich and Stanovich (2010, p. 204), this differentiation is necessary as "all hypothetical thinking involves type 2 processing [...] but not all type 2 processing involves hypothetical thinking."

#### 3.2.1.1 Two Models of CT

Based on the tripartite model of thinking, two different models can be conceived that explain the activation of CT (examples are given in the following sub-section):

(1) A one-dimensional model that comprises all activations of CT as type 2 thinking; in other words, reflective thinking is needed, whenever CT is activated.
(2) A two-dimensional model that distinguishes between two different kinds of activating CT, emphasizing hypothetical thinking as a special manifestation of type 2 thinking: (2a) Intuitive solutions—produced by autonomous thinking—are checked by algorithmic thinking, whereas (2b) conscious solutions (e.g., obtained by calculations) are critically reflected upon by reflective thinking.

The **research question** for this article is which model of CT—the one- or the two-dimensional—better fits the empirical data obtained with our test.

### 3.2.2 Matching the CT Test to the Two Models of CT

A theoretical and empirical analysis of the mathematics-specific CT test has been performed using students' solutions that were obtained in task-based interviews (Rott and Leuders 2016) as well as experts' solutions of the test items. This analysis revealed that the items can be matched to the two different modes of model (2). Consider, for example, item 7 (the bat-and-ball task, Kahneman 2011) as an example of type (2a). The intuitive answer—generated by the autonomous mind without calculations—most persons come up with is \$ 0.10. Activating type 2 thinking, however, reveals that the ball should cost \$ 0.05 and the bat \$ 1.05 to fit the given conditions.

Item 9, matches, on the other hand, cannot be explained by the same cognitive actions. An often expressed (wrong) solution for this item is 95 matches as a result of the multiplication 19 times 5. As the result of an effortful calculation, this solution is normally obtained by using algorithmic thinking. In this case, the correct solution of 91 matches can only be found by using reflective thinking (type (2b)) and realizing that in the calculation at hand, matches have been counted twice after six squares each.

Another example for this kind of task is item 5:

> ITEM 5: If you merge two pentagons that each have a side of common length at this very side, you will always get an octagon.
>
> This statement is ☐ correct / ☐ incorrect.

People have to consciously imagine this geometrical situation and often erroneously answer "correct". Hypothetical thinking is needed to image a situation in which sides of the two pentagons align in a way that the resulting figure has less than eight vertices.

Overall, five of the test items are of type (2a) like item 7, whereas the remaining nine items are of type (2b) like items 5 or 9 (see Table 3.2). Therefore, the CT test should allow for a distinction between the two types of model (2), and, thus, enable an answer to the research question regarding the one- versus two-dimensional model of CT.

### 3.2.3 Methodology

The CT test has been used in two large samples of mathematics pre-service teachers from the universities of the authors. Both times, the tasks were rated dichotomously, 1 point for a correct answer and 0 points for a wrong answer.

- Study I: At the beginning of the winter term 2013/14, $n = 150$ students completed the first version of the CT test with 11 items.
- Study II: At the beginning of the winter term 2014/15, $n = 468$ students completed a version of the CT test with 3 additional items (14 items in total).

To answer the research question regarding the dimensionality of CT, two independent methodological approaches have been pursued: (a) exploratory factor analysis (and factor extraction using varimax rotation) using the program SPSS and (b) Rasch modelling (Boone et al. 2014) using the program Winsteps (Linacre 2005).

## 3.3 Results

To examine whether the CT test is able to differentiate between high and low achieving students, a Rasch analysis has been performed. Doing so, all items (as well as all persons) are expressed on a linear scale. The mean item measure is set to 0.00 logits with easier items having negative and more difficult items positive measures (Boone et al. 2014, p. 71). The item measures (Table 3.4) show a broad distribution of easy to hard items that cover the whole scale without large gaps. This verifies the ability of the test to differentiate between high and low achieving students. Additionally, Table 3.4 shows mean values and standard deviations of the solution rates of the CT test. A comparison of these values reveals very robust solution rates across the two studies (the solution rates are also in line with those of the pilot studies, see Rott and Leuders 2016).

(a) Exploratory factor analysis to examine the dimensionality of the CT test

In the cases of both study I and II, exploratory factor analyses come to identical results: The Kaiser criterion (eigenvalues > 1, horizontal line in Fig. 3.2) suggests a five-factor solution, with a very high eigenvalue for the first factor and a strong decline in the eigenvalues of the following factors. Fittingly, Cattell's scree test

**Table 3.4** Rasch Measure (R.M.), mean values (Mean), and standard deviations (S.D.) for each item in studies I and II

| Item | 1 | 2 | 3 | 4 | 5 | 6 | 7 | 8 | 9 | 10 | 11 | 12 | 13 | 14 |
|---|---|---|---|---|---|---|---|---|---|---|---|---|---|---|
| Study I: winter term 2013/14, $n = 150$ | | | | | | | | | | | | | | |
| R.M. | −0.45 | 0.65 | −2.63 | 0.44 | 0.03 | 0.54 | 0.03 | 20.46 | −1.21 | 1.28 | −1.14 | – | – | – |
| Mean | 0.53 | 0.31 | 0.88 | 0.35 | 0.43 | 0.33 | 0.43 | 0.09 | 0.68 | 0.21 | 0.67 | – | – | – |
| S.D. | 0.50 | 0.46 | 0.33 | 0.48 | 0.50 | 0.47 | 0.50 | 0.28 | 0.47 | 0.41 | 0.47 | – | – | – |
| Study II: winter term 2014/15, $n = 468$ | | | | | | | | | | | | | | |
| R.M. | −0.76 | 0.37 | −1.97 | 0.20 | −0.19 | 0.81 | −0.55 | 1.64 | −1.21 | 1.19 | −1.34 | −0.34 | 2.00 | 0.14 |
| Mean | 0.59 | 0.36 | 0.80 | 0.39 | 0.47 | 0.28 | 0.55 | 0.16 | 0.68 | 0.22 | 0.70 | 0.50 | 0.12 | 0.40 |
| S.D. | 0.49 | 0.48 | 0.40 | 0.49 | 0.50 | 0.45 | 0.50 | 0.37 | 0.37 | 0.42 | 0.46 | 0.50 | 0.33 | 0.49 |

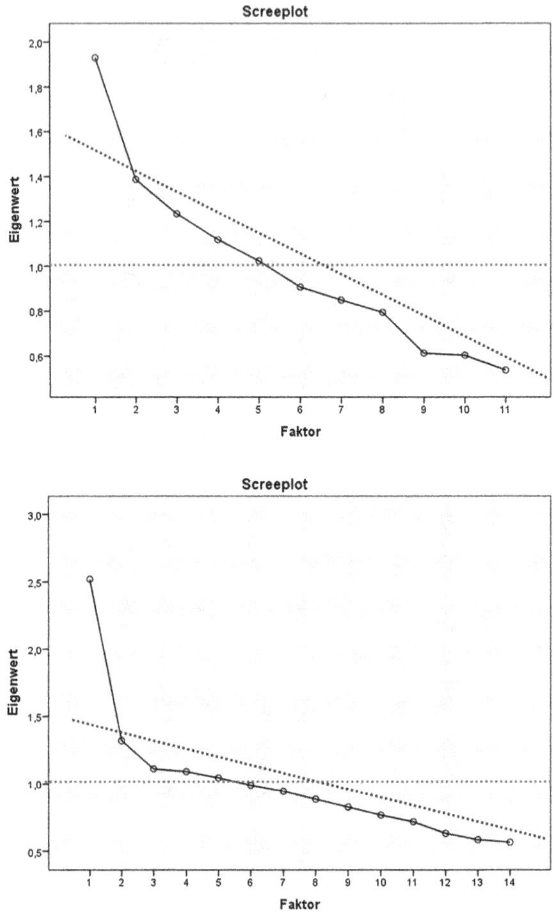


**Fig. 3.2** Scree plots of the factor analyses in studies I and II

("elbow criterion", oblique line in Fig. 3.2) suggests a one-factor solution (with two factors being possible).

Although the scree tests suggest one-factor solutions, exploratory two-factor solutions have been calculated to see whether the empirical factors match the theoretical classification of the items regarding types (2a) and (2b) (Table 3.5).

**Table 3.5** Theoretical classifications (t.c.) and factor loadings for factor 1 (f1) and factor 2 (f2) for each item in studies I and II (values greater than 0.3 in *bold*)

| Item | 1 | 2 | 3 | 4 | 5 | 6 | 7 | 8 | 9 | 10 | 11 | 12 | 13 | 14 |
|---|---|---|---|---|---|---|---|---|---|---|---|---|---|---|
| t.c. | 2b | 2a | 2b | 2a | 2b | 2a | 2a | 2b | 2b | 2b | 2b | 2b | 2b | 2a |
| Study I: winter term 2013/14, $n = 150$ | | | | | | | | | | | | | | |
| f1 | **0.55** | −0.04 | **0.35** | **0.33** | −0.03 | **0.34** | **0.69** | 0.13 | **0.55** | **0.65** | 0.15 | – | – | – |
| f2 | −0.28 | **0.62** | **−0.63** | **0.51** | −0.01 | 0.28 | −0.07 | −0.25 | −0.16 | 0.17 | **0.32** | – | – | – |
| Study II: winter term 2014/15, $n = 468$ | | | | | | | | | | | | | | |
| f1 | **0.65** | 0.05 | 0.17 | −0.14 | 0.07 | 0.17 | **0.68** | **0.51** | **0.64** | **0.44** | **0.35** | **0.61** | 0.18 | **0.39** |
| f2 | −0.10 | **0.69** | **−0.43** | **0.72** | −0.10 | 0.04 | −0.09 | 0.15 | −0.01 | −0.12 | 0.25 | −0.16 | −0.03 | −0.06 |

**Table 3.6** MNSQ-fit values for each item in studies I and II

| Study I: winter term 2013/14, $n = 150$ | | | | | | | | | | | |
|---|---|---|---|---|---|---|---|---|---|---|---|
| **Item** | **1** | **2** | **3** | **4** | **5** | **6** | **7** | **8** | **9** | **10** | **11** |
| MNSQ | .94 | 1.15 | .99 | 1.03 | 1.13 | 1.00 | .81 | 1.07 | .90 | .86 | 1.10 |
| Study II: winter term 2014/15, n = 468 | | | | | | | | | | | |
| **Item** | **1** | **2** | **3** | **4** | **5** | **6** | **7** | **8** | **9** | **10** | **11** |
| MNSQ | .86 | 1.18 | 1.06 | 1.14 | 1.18 | 1.11 | .82 | .89 | .85 | .94 | .97 |
| **Item** | **12** | **13** | **14** | | | | | | | | |
| MNSQ | .89 | 1.06 | .98 | | | | | | | | |

Except for item 3 (which shows ceiling effects in its solution rates), there are no significant negative factor loadings. The empirically found factors for both studies show no interpretable mapping to the theoretical classifications. In both studies, factor 2 contains a very small number of items that do not belong to one of the two possible modes of the two-dimensional model of CT. In study II, the second factor consists of only items 2 and 4, which are very similar to each other and also have similar solution rates. Our interpretation of the factor analyses is that there is only one factor.

(b) Rasch analysis to examine the dimensionality of the CT test

The requirement for a Rasch analysis is that all items of the test involve only one general issue. In other words, the instrument has to be one-dimensional. Conducting a Rasch analysis therefore always involves tests that check for each item (and each person) whether this requirement is violated (Boone et al. 2014). Therefore, mean squared (MNSQ) fit values are reported (see Table 3.6) to see whether the items are within acceptable ranges or have to be discarded for further analyses. Generally, a range between 0.5 and 1.5 is accepted to be a reasonable fit of the data to the model (ibid., p. 166).

None of the MNSQ-fit values are outside the range of 0.5 and 1.5, suggesting a good fit to a one-dimensional model which the test is based upon.

### 3.3.1 Discussion

Based on the tripartite model of cognition, the research question addressed in this article was whether it would be possible to empirically differentiate two dimensions of CT that belong to two processes in the cognitive model. This approach is in line with the requirements of cognitive diagnostic modelling (CDM, Rupp and Mislevy 2007).

Two independent analyses of the data of more than 600 students both suggest that mathematical critical thinking—as measured with the test at hand—is a one-dimensional construct: Regardless whether a solution has been obtain subconsciously/intuitively or consciously/by calculations, checking and critically reflecting upon the solution seems to involve the reflective mind.

Looking back at item 7 (bat-and-ball) and the intuitive solution of $ 0.10, this result seems plausible: Even though a simple addition of $ 0.10 and $ 1.10 could prove this answer to be wrong, the result has to be questioned (reflective thinking), before it can be checked by a calculation (algorithmic thinking).

This finding could have implications for all situations in which solutions have to be critically reviewed. Mere algorithmic reviews of a solution might not be sufficient if there is no disposition to engage in hypothetical (reflective) thinking.

##### 3.3.1.1 Limitations

There are, however, limitations to the results presented in this paper. Regarding the methodology used in this study, one could argue that instead of exploratory factor analysis, confirmatory factor analysis should have been used. This approach was rejected because of problems with this very method (for details, see Prudon 2015).

Other limitations regard the design of the studies: Even though the results favour a one-dimensional model, this does not mean that the construct has to be one-dimensional. This only demonstrates that the proposed operationalization cannot reflect the two dimensions that have been assumed on a theoretical basis. This may have several reasons that suggest further investigations:

*Validity of the items*: Do the operationalizations in the constructed items reflect the assumed processes or is there a mixture of solution processes that fails to differentiate between the dimensions?

*Variance in the population*: Is there a sufficient variance structure in the group under investigation or is the expected multi-dimensionality reduced by a common ability? This could be investigated by introducing experimental conditions that trigger the different processes before solving the items.

*Differential validity*: Although we tried to keep the required mathematical knowledge as low as possible, it may have an influence on the solution rates. Similarly, other general cognitive factors such as inductive reasoning may have an impact on the results. Therefore, further studies have to include moderating variables to gain knowledge about the differential validity of the proposed operationalization of CT.

Furthermore, with regard to the broad use of CT in all areas of education, the domain specificity should be an interesting area for further research.

## 3.4 Mathematical Critical Thinking: Correlations with Other Constructs

In this section, we want to shortly address possible correlations of mathematical critical thinking (measured with the test described above) with similar constructs, namely (1) *problem solving*, (2) *mathematical reasoning, argumentation, and proof*, as well as (3) *metacognition.*

### 3.4.1 Correlations Between Critical Thinking and Problem Solving

Because of its conceptualizations and definitions, CT is closely related to (mathematical) problem solving (cf. Facione 1990; Lai 2011; Jablonka 2014). Therefore, an independent test measuring problem solving ability should show significant correlations with the CT test described earlier in this chapter.

A study with $n = 76$ students from an upper secondary school in Essen was conducted (Piecyk 2017). The CT test (with an additional item) as well as a problem-solving test were used to explore this possible correlation. The latter one consisted of three problems:

> PROBLEM 1: There are 30 rows of seats in the auditorium of a theatre. There are 2 more seats in each row than in the previous one. How many seats are there if there are 50 in the 15th row?
> PROBLEM 2: Nine positive integers are given, which are arranged in such a sequence that the sum of three consecutive numbers is equal. The first number in order is 450, the last 50. The sum of all numbers is 2010. Determine all nine numbers.
> PROBLEM 3: With how many zeroes ends $1 \cdot 2 \cdot 3 \cdot \ldots \cdot 99 \cdot 100$?

For each step of solving the problems, the participants were given points, resulting in scores between 0 and 24 points for the problem-solving test. In the CT test, the students were awarded with one point for each correct solution.

A chi-square test (with Yates correction) reveals a significant correlation between both tests (Table 3.7, this evaluation was done by BR).

**Table 3.7** Comparison of the Problem Solving (PS) and Critical Thinking (CT) tests; the numbers in brackets indicate expected frequency under the assumption of statistical independence

| | | Critical thinking | | |
|---|---|---|---|---|
| | | 0–8 | 9–15 | Sum |
| PS | 0–10 | 27 (22.0) | 13 (16.0) | 38 |
| | 11–24 | 17 (22.0) | 21 (16.0) | 38 |
| | Sum | 44 | 32 | 76 |
| | | $\chi^2 = 4.37$ | df = 1 | $p = 0.036$ |

### 3.4.2 Correlations Between Critical Thinking and Mathematical Reasoning, Argumentation, and Proof

As CT encompasses judging, evaluating, analyzing arguments and claims, as well as inductive or deductive reasoning (cf. Facione 1990; Lai 2011), a correlation between CT and mathematical reasoning, argumentation, and proof is expected.

To check this expectation, a study with $n = 74$ university students (pre-service teachers of mathematics) was conducted (Blum 2017). The CT test (with 15 items) was used as well as a test with two arithmetic tasks and one geometry task that involved reasoning and proof. An ordinal categorization scheme with six categories (from 0: no processing to 5: complete proof; see Brockmann-Behnsen and Rott 2014, for details) was used to evaluate the mathematical reasoning abilities of the test persons. In this test, the students were asked to proof the statement of each task, examples of the tasks are:

> TASK 1: The sum of three consecutive natural numbers can always be divided by three.
> TASK 2: Any square number $n^2$, where $n$ is a natural number greater than 1, when divided by 4, can leave only the remainder 0 or 1.

A chi-square test (with Yates correction) reveals a significant correlation between both tests (Table 3.8, this evaluation was done by BR).

Additionally, seven task-based interviews were conducted in which students were asked to solve additional CT items as well as additional reasoning tasks. Analyses of the videotaped processes show better reasoning skills from the students that performed better in the CT items and vice versa, further confirming the hypothesis of a correlation between CT and mathematical reasoning.

**Table 3.8** Comparison of the Reasoning, Argumentation, and Proof test (RAP) and the Critical Thinking (CT) test; for the RAP, the maximum category (as a measure for reasoning ability) was selected

| | | Critical Thinking | | |
|---|---|---|---|---|
| | | 0–8 | 9–15 | Sum |
| RAP | 0–3 | 39 (32.8) | 15 (21.2) | 54 |
| | 4–5 | 6 (12.2) | 14 (7.8) | 20 |
| | Sum | 45 | 29 | 74 |
| | | $\chi^2 = 9.22$ | df = 1 | $p = 0.0024$ |

The numbers in brackets indicate expected frequency under the assumption of statistical independence

### 3.4.3 Correlations Between Critical Thinking and Metacognition

Finally, a study with 112 students (pre-service teachers of mathematics) was conducted to test for possible correlations of CT with metacognition and self-regulation, which could also be expected because of the definition of CT (cf. Facione 1990; Lai 2011).

To measure metacognition, the self-report instrument *MAI—Metacognitive Awareness Inventory* (Schraw and Dennison 1994) was used, which consists of 52 statements that are rated on a five-point Likert-scale (from 1: "totally disagree" to 5: "totally agree" with reversed polarities for some items), resulting in possible scores from 52 to 260 points with a high number of points indicating a high awareness of metacognition.

A chi-square test (with Yates correction) reveals a significant correlation between both tests (Table 3.9, this evaluation was done by BR). A Pearson correlation of the same data shows a significant correlation coefficient ($r = 0.489$, $p < 0.1$) between both test results.

Additionally, two interviews were conducted with students that scored good (above the median) in one test but bad in the other (below the median). In these two interviews, it could be confirmed that these "statistical outliers" had not read the self-report questionnaire carefully or worked on the CT test seriously, respectively.

**Table 3.9** Comparison of the Metacognition and Self-Regulation test (M/S) and the Critical Thinking (CT) test; the numbers in brackets indicate expected frequency under the assumption of statistical independence

| | | Critical Thinking | | |
|---|---|---|---|---|
| | | 0–6 | 7–15 | Sum |
| M/S | 52–175 | 43 (29.9) | 11 (24.1) | 54 |
| | 176–260 | 19 (32.1) | 39 (25.9) | 58 |
| | Sum | 62 | 50 | 112 |
| | | $\chi^2 = 23.00$ | df = 1 | $p < 0.0001$ |

### 3.4.4 Discussion of the Three Correlation Studies

The three (excerpts from) studies presented in this section explored possible connections between the CT test and the constructs problem solving, mathematical reasoning, and metacognition. For each of those constructs, theoretical considerations predicted significant statistical correlations. All three studies confirmed these correlations, which indicates the validity of the construct "mathematical critical thinking" as measured with our CT test. However, all observed correlations were far from perfect, but only moderately strong. This is interpreted in the following way: Even though CT is correlated to problem solving, mathematical reasoning, and metacognition, it is a different, independent construct.

Future studies could investigate possible correlations between CT and other constructs like intelligence (measured with IQ tests).

## References


Boone, W. J., Staver, J. R., & Yale, M. S. (2014). *Rasch Analysis in the Human Sciences.* Dordrecht: Springer.

Blum, S. (2017). *Untersuchungen zum Zusammenhang von mathematisch-kritischem Denken und mathematischem Argumentieren bei Studierenden des Lehramts.* Master Thesis, University of Duisburg-Essen.

Brockmann-Behnsen, D., & Rott, B. (2014). Fostering the Argumentative Competence by Means of a Structured Training. In S. Oesterle, C. Nicol, P. Liljedahl, & D. Allan (Eds.), *Proceedings of the Joint Meeting of PME 38 and PME-NA 36, Vol. 2* (pp. 193–200). Vancouver: PME

Cacioppo, J. T., Petty, R. E., Feinstein, J., & Jarvis, W. (1996). Dispositional differences in cognitive motivation: The life and times of individuals varying in need for cognition. *Psychological Bulletin, 119*(2), 197–253.

Elezkurtaj, D. (2017). *Untersuchungen zum Zusammenhang von mathematisch-kritischem Denken und Metakognition/Selbstregulation*. Master Thesis, University of Duisburg-Essen.

Ennis, R. H., & Weir, E. (1985). *The Ennis-Weir critical thinking essay test*. Pacific Grove: Midwest Publications.

Facione, P. A. (1990). *Critical thinking: A statement of expert consensus for purposes of educational assessment and instruction*. Executive Summary "The Delphi Report". Millbrae: The California Academic Press.

Jablonka, E. (2014). Critical thinking in mathematics education. In S. Lerman (Ed.), *Encyclopedia of Mathematics Education* (pp. 121–125). Dordrecht: Springer.

Kahneman, D. (2003). A perspective on judgment and choice. *American Psychologist, 58*, 697–720.

Kahneman, D. (2011). *Thinking, fast and slow*. London: Penguin Books Ltd.

Kaput, J. J., & Clement, J. (1979). Letter to the editor. *The Journal of Children's Mathematical Behavior, 2*(2), 208.

Lai, E. R. (2011). *Critical Thinking: A Literature Review*. Upper Saddle River: Pearson. www.pearsonassessments.com/hai/images/tmrs/criticalthinkingreviewfinal.pdf.

Linacre, J. M. (2005). Winsteps Rasch analysis software. PO Box 811322, Chicago IL 60681-1322, USA.

Montazeri, S. (2017). *Mathematisch kritisches Denken – ein Vergleich von zwei 9. Klassen an Gymnasien*. State Examination Thesis, University of Duisburg-Essen.

Pearson Education. (2012). *Watson-Glaser Critical Thinking Appraisal User-Guide and Technical Manual*. http://www.talentlens.co.uk/assets/news-and-events/watson-glaser-user-guide-and-technical-manual.pdf

Piecyk, S. (2017). *Untersuchung zum Zusammenhang von mathematisch kritischem Denken und mathematischem Problemlösen bei Schülerinnen und Schülern der Sekundarstufe II*. Master Thesis, University of Duisburg-Essen.

Prudon, P. (2015) Confirmatory factor analysis as a tool in research using questionnaires: a critique. *Comprehensive Psychology, 4*, article 10.

Raßmann, K. (2015). *Quantitative und qualitative Untersuchungen zum mathematisch-kritischen Denken*. State Examination Thesis, University of Duisburg-Essen.

Rott, B., Leuders, T., & Stahl, E. (2015). Assessment of mathematical competencies and epistemic cognition of pre-service teachers. *Zeitschrift für Psychologie, 223*(1), 39–46.

Rott, B., & Leuders, T. (2016). Mathematical critical thinking: The construction and validation of a test. In C. Csíkos, A. Rausch, & J. Szitányi (Eds.), *Proceedings of the 40th Conference of the International Group for the Psychology of Mathematics Education, Vol. 4* (pp. 139–146). Szeged: PME.

Rott, B., & Leuders, T. (2017). Mathematical Critical Thinking: A Question of Dimensionality. In B. Kaur, W. K. Ho, T. L. Toh, & B. H. Choy (Eds.), *Proceedings of the 41st Conference of the International Group for the Psychology of Mathematics Education, Vol. 1* (p. 263). Singapore: PME.

Rupp, A. A., & Mislevy, R. J. (2007). Cognitive foundations of structured item response theory models. In J. Leighton & M. Gierl (Eds.), *Cognitive diagnostic assessment in education: theory and applications* (pp. 205–241). Cambridge: Cambridge University Press.

Schraw, G., & Dennison, R. S. (1994). Assessing Metacognitive Awareness. *Contemporary Educational Psychology, 19*, 460–475.

Stanovich, K. E., & Stanovich, P. J. (2010). A framework for critical thinking, rational thinking, and intelligence. In D. Preiss & R. J. Sternberg (Eds.), *Innovations in educational psychology: Perspectives on learning, teaching and human development* (pp. 195–237). New York: Springer.

Weinert, F. E. (2001). Concept of competence: A conceptual classification. In D. S. Rychen & L. H. Salganik (Eds.), *Defining and selecting key competencies*. Göttingen: Hogrefe.

# 4 Critical Thinking and Epistemological Beliefs of Pre-service Teachers

As a continuation and further development of the interview study presented in Chap. 2, a questionnaire was developed to be able to assess epistemological beliefs of a large number of participants in general or of pre-service teachers in this specific study, respectively. One goal was to see whether the different arguments given in the interviews could be found within a population solely consisting of students (without experts like professors of mathematics).

This chapter is based on a (peer-reviewed) journal article (Rott et al. 2015b).[1] Because of space limits, the original article could not contain extensive information regarding interrater agreements of our instruments or contingency tables of statistical tests. This information is added in this chapter to document the rigor of our work.

## 4.1 Background

Higher education is a key to the development of modern society, especially in disciplines like teacher education, because these disciplines deal with the transfer of knowledge and are supposed to have a leverage effect. While the measurement of competencies in primary and secondary education has evolved considerably during the past decades (Hartig et al., 2008), there is still a need for competence

[1] The article was published in the journal *Zeitschrift für Psychologie* by Hogrefe Publishing (doi: https://doi.org/10.1027/2151-2604/a000198). In their terms of use, Hogrefe allows for accepted manuscripts (after review) "as part of a grant application or submission of a thesis or doctorate: may be shared at any time." This version of the article may not completely replicate the final version published in [journal title]. It is not the version of record and is therefore not suitable for citation.



B. Rott, *Epistemological Beliefs and Critical Thinking in Mathematics*, Freiburger Empirische Forschung in der Mathematikdidaktik, https://doi.org/10.1007/978-3-658-33539-7_4

| | **Mathematical thinking** | **epistemological beliefs on mathematical knowledge and knowing** | |
|---|---|---|---|
| **high level** | critical thinking | sophisticated beliefs | |
| ↑ | ↑ | ↑ | denotative judgments |
| **low level** | algorithmic thinking | inflexible beliefs | |
| | autonomous mind | connotative beliefs | |

**Fig. 4.1** Dimensions and development of mathematical competencies

models for assessment in higher education (Blömeke et al. 2013a, b). Although instruments for assessing competencies in teacher education on a large scale are available (Blömeke et al. 2013a, b; Baumert et al. 2010; Hill et al. 2005; OECD 2010), many questions on the underlying competence structures remain open.

Competencies are considered as complex abilities closely related to performance in real-life-situations (Hartig et al. 2008; Shavelson 2010; Blömeke et al. 2015) and they comprise and integrate constructs such as skills, abilities, knowledge, beliefs, and motivation (Weinert 2001). Competence models are expected to specify the structure of competencies in pre-defined areas and constitute a framework for validly measuring competencies. Within the specific area of mathematics, teacher education it has been studied how competencies develop and how abilities and beliefs interconnect (e.g., Staub and Stern 2002; Schoenfeld 2003); however, there is still a lack of approaches to analyze these competence structures by means of psychometric competence models. Therefore, in our study (which is embedded in a larger initiative "Modeling and Measuring Competencies in Higher Education", cf. Blömeke et al. 2013a, b) we construct and use a competence model which incorporates epistemological beliefs about mathematics and critical thinking in mathematics as two central cognitive dimensions of pre-service mathematics teachers. Figure 4.1 presents an overview on the relevant constructs and variables of the study explained in the following sections.

### 4.1.1 Structure of the Belief Dimension

Epistemological beliefs can be defined as learners' beliefs about the nature of knowledge and knowing (Hofer and Pintrich 1997; Hofer 2000). A growing amount of empirical evidence shows that epistemological beliefs are related to several aspects of learning processes and to learning outcomes (e.g., Buehl and Alexander 2006; Hofer and Pintrich 1997). Furthermore, students' epistemological beliefs are affected by their teachers' epistemological beliefs and their teaching style (e.g., Brownlee and Berthelsen 2008; Haerle and Bendixen 2008). Adequate epistemological beliefs are considered as a prerequisite to successfully complete higher education (e.g., Trautwein and Lüdtke 2004; Bromme 2005) and for an elaborated understanding of scientific findings. Beliefs of the nature of knowledge are regarded as a prerequisite for an active civic participation in modern science- and technology-based societies (Bromme 2005). However, they are rarely included in competence models, as for example in the area of mastering languages (cf. Klieme et al. 2004, p. 135 ff.). In our study we wish to introduce a model which incorporates beliefs on the nature of knowledge in the area of mathematics.

To empirically assess beliefs, one can draw on various theories on their structure:

(a) Epistemological beliefs are usually seen as *multidimensional*. A widely accepted structure was proposed by Hofer and Pintrich (1997), who differentiated between two general areas of epistemological beliefs (*nature of knowledge* and *nature or process of knowing*) with two dimensions each (*certainty* and *simplicity* as well as *source* and *justification of knowledge*). For the aim of this study, we focus on beliefs about the *certainty of knowledge*, which is central to mathematics (Kline 1980). In the public opinion, mathematics is regarded as a domain of certain knowledge and many misconceptions on mathematics are related to this view (Muis 2004; Weber et al. 2014).

(b) Most theoretical approaches include different *levels of specificity* of epistemological beliefs. Hofer (2006) distinguishes between general epistemological beliefs, disciplinary perspectives on beliefs and discipline-specific beliefs. There is strong evidence for the assumption that epistemological beliefs are context-related (e.g., Stahl 2011). Hence, in our study we examine interactions between mathematical knowledge and beliefs about mathematics on a discipline-specific level. To further account for the context specificity of epistemological beliefs we distinguish between beliefs about mathematics expressed in either a connotative or a denotative way. Connotative judgments are defined as judgments about the nature of mathematics that are activated spontaneously when no further context is given; for example, when a student is asked whether he or she generally thinks that mathematical knowledge is rather certain or uncertain. It is assumed that these judgments are directly related to the discipline-specific epistemological beliefs. Denotative judgments are generated with respect to a specific (mathematical) situation and are expected to be more reflected and more context-specific (Bromme et al. 2008); e.g., when a student is asked to choose between two given positions and give arguments for his choice.

(c) Nearly all existing approaches address the *development* of beliefs during education towards more sophisticated epistemological beliefs. Often it is assumed that a strong absolutistic view on knowledge, which stresses its certainty and stability and considers knowledge as accumulation of facts that can be transferred by authority, is not considered appropriate in many contexts in which individuals have to judge knowledge claims. On the other hand, a strong relativistic view on knowledge that stresses its uncertainty and instability and that might result in an acceptance of different viewpoints without deeper reflections is also not appropriate for reflected epistemological judgments (Krausz 2010). Therefore, we agree with Bromme et al. (2008) who describe sophisticated epistemological beliefs "as those beliefs which allow for context-sensitive judgments about knowledge claims".[2] This is in line with an evaluativistic perspective (e.g., King and Kitchener 2002) which accepts a certain degree of uncertainty and changeability of truth. This also goes well with the fact that academic disciplines have commonly accepted methods and standards in the development of scientific knowledge. Therefore, aspects like the ontology of a discipline (Bromme et al. 2008), the specific context (e.g., Elby and Hammer 2001), and the sociocultural context (e.g., Buehl and Alexander 2006) should be taken into account, when knowledge claims are judged. This view of sophistication as an epistemologically reflected way to deal with knowledge claims supports the idea of Stahl (2011) that it is necessary to distinguish between the epistemological orientation (e.g., more absolutistic, more relativistic, more evaluativistic) and the sophistication of the judgment (level of reflection).

### 4.1.2 Structure of the Knowledge Dimension

When assessing knowledge (as the second cognitive dimension of our competence model) we do not measure mathematical achievement in terms of students' knowledge of the content of university courses. A theoretically more coherent picture of the students' competence can be achieved by capturing the quality of the *use* of mathematical knowledge by drawing on the concept of *critical thinking*. Facione (1990, p. 3) "understand[s] critical thinking to be purposeful, self-regulatory judgment which results in interpretation, analysis, evaluation, and inference, as well as

[2] Please note that we do not intend to use "sophistication" to convey a value judgment (cf. Muis 2004, p. 332 f.), but rather to indicate a reflected use of arguments.

explanation of the evidential, conceptual, methodological, criteriological, or contextual considerations upon which that judgment is based. […].” Though many different conceptualizations of critical thinking exist (e.g., in philosophy, psychology, and education) the following abilities are commonly agreed upon (cf. Lai 2011, p. 9 f.): analyzing arguments, claims, or evidence; making inferences using inductive or deductive reasoning; judging or evaluation and making decisions; or solving problems.

To locate critical thinking within cognition, Stanovich and Stanovich (2010) propose a tripartite model of thinking, adapting and extending *dual process theory* (e.g., Kahneman 2003). They distinguish the subconscious thinking of an “autonomous mind” from the conscious thinking of an “algorithmic mind” and a “reflective mind” (see Fig. 4.2). Critical thinking is identified with the functioning of the reflective mind, which may override algorithmic processes. When solving mathematical problems critical thinking can be attributed to those processes that consciously regulate the algorithmic use of mathematical procedures. A context-specific operationalization of critical thinking can be considered as a competence component (Weinert 2001).

Consequently, tasks to measure critical thinking that reflect this definition should (i) reflect discipline specific solution processes but should *not* require higher level mathematics, (ii) require a reflective component of reasoning and judgment when solving a task or evaluating the solution, (iii) reflect an appropriate variation of difficulty within the population.

Resting upon these conceptualizations of the belief and knowledge dimensions of mathematical competence, we intend to assess the levels of sophistication within the competencies of future mathematic teachers and their development at different stages of their university education:

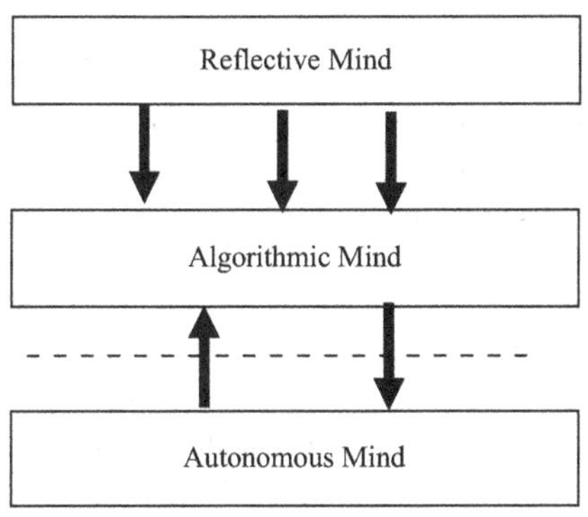


**Fig. 4.2** The tripartite model of thinking by Stanovich and Stanovich (2010, p. 210); the broken horizontal line represents the key distinction in dual process theory

**Hypothesis 1** We expect students at the beginning of their mathematical education to show lower levels of mathematical thinking and more inflexible beliefs. This finding would be coherent with the picture of mathematics conveyed in school (Schoenfeld 1983, 1989).

**Hypothesis 2** We expect that we can distinguish between students' epistemological orientations (*certain* vs. *uncertain*) on the one hand and the sophistication of their argumentations on the other hand reflecting the reported discussion on context specificity of epistemological judgments. This structure should be seen more clearly in 4th semester students than in 1st semester students, due to the influence of reflective elements of courses in mathematics and mathematics education.

**Hypothesis 3** Finally, we expect that this deeper reflection of the discipline induces connotative beliefs to be more in line with denotative beliefs in higher semesters.

By these analyses, we intend to find indications that our instruments are sensitive to identify competence profiles and changes in competencies, also when used within long-term longitudinal settings in subsequent studies.

## 4.2 The Study

Measuring the impact of university education on beliefs and critical thinking may pose severe problems. Arum and Roksa (2011) report that most students do not improve their critical thinking skill during their first two years of university studies. The CBMS (2012, p. 55) summarizes the state of mathematics teacher education in the United States (which is comparable to Germany): "A primary goal of a mathematics major program is the development of mathematical reasoning skills. This may seem like a truism to higher education mathematics faculty, to whom reasoning is second nature. But precisely because it is second nature, it is often not made explicit in undergraduate mathematics courses." Therefore, it may prove difficult to assess the change of students' beliefs or higher order skills with any instrument in regular teacher education. However, we situated our research within the teacher education program of the "University of Education Freiburg" which specializes on teacher education and only appoints specialists in education, in educational as well as in all subject matter courses. During the first four semesters, future mathematics teachers attend courses on education, mathematics education and mathematics. All courses comprise reflective elements on the methods and the epistemology of mathematics as a discipline, e.g. the role of proof on different levels of rigor, experimentation in mathematics, etc. (Barzel et al. 2016). Therefore, we expect to find relevant growth in competence areas that address the level of reflection on mathematical processes and meta-mathematical issues.

Participants of the study were students at the University of Education Freiburg with the aim of teaching mathematics in primary and secondary schools. In 2013/14, two groups were observed longitudinally at two times of measurement each: Group 1 at the beginning and end of the 1st semester ($T_1$, $n = 105$; $T_2$, $n = 87$; full survey on all beginning students) and group 2 at the beginning and end of the 4th semester ($T_3$, $n = 42$; $T_4$, $n = 59$; participants of a seminar on epistemological processes in mathematics). At $T_1$ and $T_3$, we collected data regarding (i) discipline-specific denotative epistemological beliefs focusing on *certainty of mathematical knowledge*, (ii) discipline-related connotative epistemological beliefs, and (iii) critical thinking in mathematics. The denotative beliefs have additionally been gathered at $T_2$ and $T_4$ to capture changes of beliefs.

## 4.2.1 Methodology

### 4.2.1.1 Measuring Epistemological Beliefs (Denotative Versus Connotative)

To gain access to denotative epistemological beliefs and to the sophistication of the students' judgments, we conducted a preliminary interview study (Rott et al. 2014, 2015a, see also Chap. 2). The interviewees' answers revealed a broad spectrum of possible responses that helped us to construct a web-based questionnaire. In this questionnaire, we used open-ended questions and prompts explaining controversial points of view towards mathematics as a scientific discipline to acquire a vivid account of our participants' epistemological judgments and according arguments. We developed a coding manual for the belief orientation (mathematics as *certain* vs. *uncertain*) and the level of argumentation (as *inflexible* vs. *sophisticated*). Overall, 293 responses (from the participants of all four measurement times) have been independently coded by two trained raters.

Table 4.1 shows that the two raters were able to code certain/ uncertain with absolute accuracy and that the overall interrater agreement was high (average Cohen's $\kappa = 0.88$).

To measure the connotative epistemological beliefs, we used the CAEB (Connotative Aspects of Epistemological Beliefs) questionnaire by Stahl and Bromme (2007) which can be used with relation to different disciplines or contexts. It consists of 24 pairs of contrastive adjectives like *simple* vs. *complex*. The respondents

**Table 4.1** Interrater agreement (between R1 and R2; R1 being the author of this book) for coding denotative beliefs. All students at all times of measurement

| | | R1 | | | | |
|---|---|---|---|---|---|---|
| | | Certain & inflexible | Certain & sophisticated | Uncertain & inflexible | Uncertain & sophisticated | Sum |
| R2 | Certain & inflexible | 89 | 6 | 0 | 0 | **95** |
| | Certain & sophisticated | 3 | 6 | 0 | 0 | **9** |
| | Uncertain & inflexible | 0 | 0 | 149 | 7 | **156** |
| | Uncertain & sophisticated | 0 | 0 | 6 | 27 | **33** |
| | Sum | **92** | **12** | **155** | **34** | **293** |
| | $P_{obs} = 0.925$, $P_{exp} = 0.398$, Cohen's $\kappa = 0.88$ | | | | | |

are supposed to judge the character of a discipline on a 7-point Likert scale for each contrastive pair. Stahl and Bromme (2007) validated the instrument in two studies with more than 1000 participants each and identified two factors via factor analysis: *Texture* (beliefs about the structure and accuracy of knowledge) and *Variability* (beliefs about the stability and dynamics of knowledge).

In the present study, we instructed our participants to complete the CAEB with "mathematics as a scientific discipline" in mind. Due to our focus on epistemological judgements about the certainty of mathematics we constructed one factor focussing on certainty—based on expert ratings and on an exploratory factor analysis. The factor consists of 10 items (Cronbach's $\alpha = 0.71$, see Table 4.2 for the items and factor loadings).

#### 4.2.1.2 Measuring Critical Thinking

Within the tripartite model of thinking (Fig. 4.2), critical thinking can be operationalized by situations that demand a critical override of algorithmic mathematical solutions by reflective and evaluative processes, such as in the paradigmatic bat-and-ball task by Kahneman and Frederick (2002): "A bat and a ball cost \$ 1.10 in total. The bat costs \$ 1 more than the ball. How much does the ball cost?" The spontaneous, algorithmically produced answer that most people come up with is \$ 0.10. A critical thinker would question this answer and realize that the ball should cost \$ 0.05, whereas people that do not use critical thinking do not evaluate their first thought and adapt their solution.

By adapting and developing more than 20 items of similar character within the domain of mathematics, we constructed a test to measure the students' critical

**Table 4.2** Adjective pairs and according factor loadings for the CAEB dimension "Certainty"

| Item<br>*Mathematics as a discipline is....* | Factor loading |
|---|---|
| precise—imprecise | 0.77 |
| exact—vague | 0.77 |
| absolute—relative | 0.74 |
| sorted—unsorted | 0.71 |
| certain—uncertain | 0.70 |
| temporary—everlasting | – 0.60 |
| definite—ambiguous | 0.53 |
| stable—unstable | 0.50 |
| confirmable—unconfirmable | 0.45 |
| accepted—disputed | 0.38 |

thinking ability (for examples, see Table 4.8 in the appendix; for details, see Chap. 3). All items were related to mathematical situations and were located on a curricular level that demanded only knowledge from lower secondary education. After validating the items in a preliminary study, the final test consisted of 11 items that were rated dichotomously.

In the present study, we used a Rasch model to transform our students' test scores into values on a one-dimensional competence scale (software RUMM 2030 by Andrich et al. 2009). After eliminating two items because of underdiscrimination (fit residual > 2.5) in connection with floor and ceiling effects respectively, for each item the model showed good fit residuals (all values between –2.5 and 2.5) and no significant differences between the observed overall performance of each trait-group and its expected performance (overall-$\chi^2 = 36.2$; $df = 27$; $p = 0.11$).

## 4.2.2 Results

### 4.2.2.1 Denotative Beliefs and Sophistication

The development of the students' denotative epistemological beliefs and their degree of sophistication are summarized in Table 4.3. At the beginning of their university studies ($T_1$), more than two thirds of the students regard mathematical knowledge as uncertain (cf. Hypothesis 1 from above). This is slightly surprising and does not conform to the finding that mathematics as taught in school shapes a

**Table 4.3** Distribution of the students' denotative epistemological beliefs and their degree of sophistication

| | Measurement Time | Certain & inflexible | Certain & sophisticated | Uncertain & inflexible | Uncertain & sophisticated | Sum |
|---|---|---|---|---|---|---|
| Group 1 (1st Semester) | ($T_1$) Beginning | 29 (27.6%) | 3 (2.9%) | 66 (62.9%) | 7 (6.7%) | 105 (100%) |
| | ($T_2$) End | 48 (55.2%) | 2 (2.3%) | 29 (33.3%) | 8 (9.2%) | 87 (100%) |
| Group 2 (4th Semester) | ($T_3$) Beginning | 8 (19.0%) | 3 (7.1%) | 25 (59.5%) | 6 (14.3%) | 42 (100%) |
| | ($T_4$) End | 7 (11.9%) | 4 (6.8%) | 35 (59.3%) | 13 (22.0%) | 59 (100%) |

view of mathematics as a collection of static and reliable procedures (Schoenfeld 1989; Muis 2004). Nevertheless, at the end of their 1st semester ($T_2$), the majority of our students tend to see mathematical knowledge as certain. Because of the students' responses in the questionnaire, we attribute this change of judgment (which is significant; McNemar-test: $\chi^2 = 7.2$, $p = 0.012$, see Table 4.4) mostly to the lectures the students attended in their 1st semester, which paid a considerable attention to mathematical reasoning and proof (e.g., geometrical proofs or mathematical induction).

Of the 4th semester students ($T_3$ and $T_4$), the majority regard mathematical knowledge as uncertain. In their responses, most of them argue with deficiencies of the review procedure of mathematical publications or the many unsolved mysteries regarding prime numbers (e.g., the existence of an infinite number of prime twins, Goldbach's conjecture, etc.).

**Table 4.4** Development of the students' epistemological beliefs in their first semester

| | | ($T_2$) End | | |
|---|---|---|---|---|
| | | Certain | Uncertain | Sum |
| ($T_1$) Beginning | Certain | 15 | 4 | 19 |
| | Uncertain | 16 | 21 | 37 |
| | Sum | 31 | 25 | 56 |
| | Confidence interval | | *up* 11.96 | *low* 1.34 |

**Table 4.5** Comparison of both dimensions of the belief questionnaire for all students

| | Inflexible | Sophisticated | Sum |
|---|---|---|---|
| Certain | 37 (37.4) | 6 (5.6) | 43 |
| Uncertain | 91 (90.6) | 13 (13.4) | 104 |
| Sum | 128 | 19 | 147 |
| | $\chi^2 = 0.06$ | df = 1 | $p = 0.807$ |

Note: The numbers in brackets indicate expected frequency under the assumption of statistical independence

With respect to the degree of the students' sophistication (*inflexible/ sophisticated*) one can detect an increase of the proportion of students that use sophisticated arguments ($T_1$: 9.5%; $T_2$: 11.5%; $T_3$: 21.4%; $T_4$: 28.8%). As expected, our data shows that the two test dimensions—*certainty* and *sophistication*—are unrelated (testing all groups at $T_1$ and $T_3$ on independence yields: $\chi^2 = 0.06$, $p = 0.81$, cf. Hypothesis 2, see Table 4.5).

#### 4.2.2.2 Connotative Beliefs

The certainty dimension of the connotative beliefs measured by the CAEB instrument provides metrical data that ranges from 1 to 7 (Table 4.6). In what follows, the data regarding the connotative beliefs and critical thinking scores will each be sorted by both judgment and sophistication.

Does the CAEB data fit to the denotative belief orientations? Low CAEB scores indicate connotative beliefs of certainty and indeed students that articulated certain beliefs in the denotative questionnaire have lower CAEB scores (2nd and 3rd row in Table 4.6). This effect is especially visible for 4th semester students. However, a two-way ANOVA does not show significant effects for the group (1st/

**Table 4.6** Means (standard deviations) of groups regarding connotative epistemological beliefs (CAEB dim. "certainty")

| | Certain | Uncertain | Inflexible | Sophisticated | Total |
|---|---|---|---|---|---|
| 1st Semester ($T_1$) | 3.20 (1.29) | 3.25 (1.09) | 3.32 (1.10) | 2.44 (1.34) | 3.23 (1.15) |
| | $n = 32$ | $n = 73$ | $n = 95$ | $n = 10$ | $n = 105$ |
| 4th Semester ($T_3$) | 2.90 (1.02) | 3.77 (1.34) | 3.47 (1.31) | 3.80 (1.35) | 3.54 (1.31) |
| | $n = 11$ | $n = 31$ | $n = 33$ | $n = 9$ | $n = 42$ |
| All students combined | 3.13 (1.23) | 3.40 (1.19) | 3.35 (1.15) | 3.08 (1.48) | 3.32 (1.20) |
| | $n = 43$ | $n = 104$ | $n = 128$ | $n = 19$ | $n = 147$ |

4th semester; $F = 0.20$; $p = 0.652$) or the students' denotative judgments (*certain/ uncertain*; $F = 3.54$; $p = 0.062$) and no interaction effect ($F = 2.88$; $p = 0.092$) indicating the disparity of connotative and denotative judgments.

Do the connotative beliefs correlate with the quality of the argumentation? As expected, there are no differences in the connotative beliefs between students that argue inflexibly compared to those that argue sophisticatedly ($t = 0.93$; df = 145; $p = 0.354$) showing that both dimensions should be distinguished.

For more information on results obtained with the CAEB, see Rott et al. (2017). In the main part of this book, this instrument will not be elaborated on.

#### 4.2.2.3 Critical Thinking

The Rasch model of the students' critical thinking ability provides metrical latent variables ranging from –2.83 to 2.81 with low values indicating a low ability (see Table 4.7).

Does critical thinking depend on the denotative judgment? A two-way ANOVA has been used to investigate possible differences between the students' critical thinking scores sorted by their judgment: The 4th semester students show higher ability scores than 1st semester students ($F = 9.54$; $p = 0.002$); there is no significant effect between students regarding mathematical knowledge as "certain" compared to students regarding it as "uncertain" ($F = 1.04$; $p = 0.310$), and no interaction effect ($F = 0.02$; $p = 0.886$) which is in accordance with our expectations.

Does critical thinking correlate with the degree of sophistication? Another two-way ANOVA has been used to analyze the critical thinking scores sorted by the quality of the argumentation. The group effect is confirmed ($F = 9.54$; $p = 0.002$). As expected, students arguing sophisticatedly show higher ability scores than students arguing in an inflexible way ($F = 4.76$; $p = 0.031$). Thus, there is a marked connection between sophistication of beliefs and the ability of thinking critically when solving tasks. There is no interaction effect ($F = 1.60$; $p = 0.208$), even though the mean difference is much more prominent in the 4th semester.

### 4.2.3 Discussion

Our aim was to better understand the complex structure of competencies in higher education. We approached this question by focusing on pre-service mathematics teachers and on two cognitive dimensions—epistemological beliefs and mathematical knowledge. We especially looked at denotative beliefs about the certainty of mathematical knowledge, the sophistication of the students' argumentation and

**Table 4.7** Means (and standard deviations) of mathematical critical thinking

| | Certain | Uncertain | Inflexible | Sophisticated | Total |
|---|---|---|---|---|---|
| 1st Semester ($T_1$) | − 0.327 (0.885) | − 0.488 (0.840) | − 0.459 (0.836) | − 0.255 (1.033) | − 0.439 (0.853) |
| | $n = 32$ | $n = 73$ | $n = 95$ | $n = 10$ | $n = 105$ |
| 4th Semester ($T_3$) | 0.268 (1.295) | 0.054 (0.887) | − 0.054 (0.987) | 0.711 (0.826) | 0.110 (0.997) |
| | $n = 11$ | $n = 31$ | $n = 33$ | $n = 9$ | $n = 42$ |
| All students combined | − 0.175 (1.023) | − 0.327 (0.886) | − 0.354 (0.891) | 0.203 (1.041) | − 0.282 (0.927) |
| | $n = 43$ | $n = 104$ | $n = 128$ | $n = 19$ | $n = 147$ |

their relations to connotative beliefs and critical thinking ability in 1st and 4th semester students. We were able to construct and adapt instruments for measuring these cognitive dimensions:

(1) To capture the denotative component of beliefs we did not construct a scale but used specific contextual information (texts on certainty) to elicit a reflected elaborate judgment.[3] These judgments could be evaluated reliably by a rating procedure that distinguished not only between the direction of the judgment but also its level of sophistication.
(2) Furthermore, we constructed a reliable scale for measuring the connotative dimension of beliefs focusing on certainty by relying on the more general framework of the CAEB instrument.
(3) The dimension of mathematical ability could be operationalized within the general framework of critical thinking. The availability of such an instrument (instead of using more general tests of critical thinking, such as Ennis and Weir 1985) is pivotal for keeping the theoretical focus of our investigation on the context of mathematics.

Our inspections of the connection between the beliefs captured by connotative and denotative judgments revealed that these perspectives should be regarded as disparate. Whether this result generalizes to other areas than the certainty of mathematical knowledge (such as the epistemological status of mathematical knowledge: constructed vs. discovered) remains an open question.

The comparison between students of the 1st and 4th semester showed that the instruments are sensitive to the increase in sophistication in cognitive dimensions: The level of critical thinking when solving mathematical problems grows alongside the sophistication of (one facet of) the belief system; this finding goes along with research on problem solving (e.g., Schoenfeld 1985). However, the proportion of students showing sophisticated beliefs is rather low, which calls for optimizing the instrument to reveal more differentiated levels.

We could also find evidence for the independence of denotative beliefs and the sophistication of according arguments. Most models of epistemological development predict a change from inflexible beliefs of certainty to sophisticated beliefs of uncertainty of knowledge during education in schools and universities. Our

[3] In these texts, we used reflections on the role of proof for certainty. Please note that the German word "Beweis" used in the study has the sole meaning of "proof" as "rigorous deduction". There is no indication that our students confused this with other concepts of "proof" like "evidence by a series of examples" (for this, the German word "prüfen" would be used).

data shows that there are sophisticated representatives of "certain knowledge" as well as unreflected representatives of "uncertain knowledge". We propose to deliberately distinguish between beliefs and the sophistication of their representation; cf. Greene and Yu (2014) for a similar critique on models of epistemic cognition.

#### 4.2.3.1 Limitations and Further Studies

There are several limitations to this study, which can be ameliorated methodologically in further investigations. For example, reliably rating the judgment and sophistication of denotative belief is dependent on the students' willingness to fill out our questionnaire and write an argumentation. Also, although the items of the critical thinking test show face validity with respect to the theoretical model, a test of differential validity of this scale with respect to mathematical knowledge has yet to be undertaken.

There are also limitations regarding the selection of our sample as it only consists of students from a university specializing in teacher education. We do not assume that the observed development of cognitive competence dimensions holds true for teacher students in general. We plan to take a considerably larger sample from different universities in Germany in a subsequent study.

Even within the University of Education Freiburg, the sample is partially biased, as we assessed all students within the 1st semester but only participants of one particular seminar from the 4th semester. Although the latter group should also cover a full sample as all students have to attend this special seminar in the course of their studies. However, due to flexible study regulations, not all students attend this seminar in their 4th semester, which explains the significantly lower number of that group in this study. Further studies will take this into account by an improved longitudinal design.

Analyzing students of different semesters on the basis of the theoretical competence model revealed some interesting facts on the development of cognitive competence dimensions in higher education. Since the theoretical framework used represents a rather domain specific approach (focusing on mathematics teacher students and on the certainty of mathematics), we cannot draw conclusions as to the generalizability of our findings to other domains or groups. Nevertheless, we see an opportunity to connect our constructs to broader competence models in further research, aiming to answer questions like these: Can we distinguish the competence structure or development in mathematics of different groups of students with different professional goals (such as mathematics, teaching or engineering)? This suggests that the instruments should be refined and further validated, for example in experimental designs, allowing for the identification of factors that strongly influence competence development.

## 4.3 Appendix: Sample Items of the Mathematical Critical Thinking Test

A low Rasch location indicates an easy item; a low Rasch fit residual indicates a high discriminatory power (see Table 4.8).

**Table 4.8** Sample items of the mathematical critical thinking test

| Item in the critical thinking test | Empirical frequency of solutions | Rasch location (fit residual) |
|---|---|---|
| In a gamble, a regular six-sided die with four green faces and two red faces is rolled 20 times. You win € 25 if a certain sequence of results is shown.<br>Which sequence would you bet on?<br>☐ RGRRR ☐ GRGRRR ☐ GRRRRR | 0.21 | 1.24 (–1.064) |
| If the sum of the digits of an integer is divisible by three, then it cannot be a prime number.<br>This statement is<br>☐ correct ☐ incorrect | 0.31 | 0.537 (0.855) |
| A table-tennis bat and a ball cost € 10.20 in total. The bat costs € 10 more than the ball. How much does the ball cost? | 0.43 | 0.032 (–1.665) |
| A sequence of 6 squares made of matches consists of 19 matches (see the figure). How many matches does a sequence of 30 squares consist of? | 0.68 | – 1.152 (–0.227) |

Note: A low Rasch location indicates an easy item; a low Rasch fit residual indicates a high discriminatory power

## References

Andrich, D., Sheridan, B. E., & Luo, G. (2009). *RUMM2030: Rasch unidimensional models for measurement*. Perth: RUMM Laboratory.

Arum, R., & Roksa, J. (2011). *Academically adrift. Limited Learning on college campuses*. Chicago: The University of Chicago Press.

Barzel, B., Eichler, A., Holzäpfel, L., Leuders, T., Maaß, K., & Wittmann, G. (2016). *Vernetzte Kompetenzen statt träges Wissen – Ein Studienmodell zur konsequenten Vernetzung von Fachwissenschaft, Fachdidaktik und Schulpraxis*. Berlin: Springer.

Baumert, J., Kunter, M., Blum, W., Brunner, M., Voss, T., Jordan, A., et al. (2010). Teachers' mathematical knowledge, cognitive activation in the classroom, and student progress. *American Educational Research Journal, 47,* 133–180.

Blömeke, S., Suhl, U., & Döhrmann, M. (2013a). Assessing strengths and weaknesses of teacher knowledge in Asia, Eastern Europe and Western countries: Differential item functioning in TEDS-M. *International Journal of Science and Mathematics Education, 11,* 795–817.

Blömeke, S., Zlatkin-Troitschanskaia, O., Kuhn, C., & Fege, J. (Eds.). (2013b). *Modeling and measuring competencies in higher education*. Rotterdam: Sense.

Blömeke, S., Gustafsson, J.-E., & Shavelson, R. (2015). Beyond dichotomies: Competence viewed as a continuum. *Zeitschrift für Psychologie, 223*(1), 3–13.

Bromme, R. (2005). Thinking and knowing about knowledge—A plea for critical remarks on psychological research programs on epistemological beliefs. In M. Hoffmann, J. Lenhard, & F. Seeger (Eds.), *Activity and sign—Grounding mathematics education* (pp. 191–201). Springer.

Bromme, R., Kienhues, D., & Stahl, E. (2008). Knowledge and epistemological beliefs: An intimate but complicate relationship. In M. S. Khine (Ed.), *Knowing, knowledge and beliefs. Epistemological studies across diverse cultures* (pp. 423–441). New York: Springer.

Brownlee, J. & Berthelsen, D. (2008). Developing relational epistemology through relational pedagogy. In M. S. Khine (Ed.), *Knowing, knowledge and beliefs. Epistemological studies across diverse cultures* (pp. 405–422). New York: Springer.

Buehl, M. M., & Alexander, P. A. (2006). Examining the dual nature of epistemological beliefs. *International Journal of Educational Research, 45,* 28–42.

CBMS—Conference Board of the Mathematical Sciences. (2012). *The mathematical education of teachers II*. Providence: American Mathematical Society.

Elby, A., & Hammer, D. (2001). On the substance of sophisticated epistemology. *Science Education, 85,* 554–567.

Ennis, R. H., & Weir, E. (1985). *The Ennis-Weir critical thinking essay test*. Pacific Grove: Midwest Publications.

Facione, P. A. (1990). *Critical thinking: A statement of expert consensus for purposes of educational assessment and instruction [Executive Summary "The Delphi Report"]*. Millbrae: The California Academic Press.

Greene, J. A., & Yu, S. B. (2014). Modeling and measuring epistemic cognition: A qualitative re-investigation. *Contemporary Educational Psychology, 39,* 12–28.

Haerle, F. C., & Bendixen, L. D. (2008). Personal epistemology in elementary classrooms: A conceptual comparison of Germany and the United States and a guide for future cross-cultural research. In M. S. Khine (Ed.), *Knowing, knowledge and beliefs. Epistemological studies across diverse cultures* (pp. 165–190). New York: Springer.
Hartig, J., Klieme, E., & Leutner, D. (Eds.). (2008). *Assessment of competencies in educational contexts: State of the art and future prospects*. Göttingen: Hogrefe & Huber.
Hill, H. C., Rowan, B., & Loewenberg Ball, D. (2005). Effects of teachers' mathematical knowledge for teaching on student achievement. *American Educational Research Journal, 42*(2), 371–406.
Hofer, B. K. (2000). Dimensionality and disciplinary differences in personal epistemology. *Contemporary Educational Psychology, 25*, 378–405.
Hofer, B. K. (2006). Beliefs about knowledge and knowing: Domain specificity and generality. *Educational Psychology Review, 18*, 67–76.
Hofer, B. K. & Pintrich, P. R. (1997). The development of epistemological theories: Beliefs about knowledge and knowing and their relation to learning. *Review of Educational Research 1997, 67*(1), 88–140.
Kahneman, D. (2003). A perspective on judgment and choice. *American Psychologist, 58*, 697–720.
Kahneman, D., & Frederick, S. (2002). Representativeness revisited: Attribute substitution in intuitive judgment. In T. Gilovich, D. Griffin, & D. Kahneman (Eds.), *Heuristics and biases: The psychology of intuitive judgment* (pp. 49–81). New York: Cambridge University Press.
King, P. M., & Kitchener, K. S. (2002). The reflective judgment model: Twenty years of research on epistemic cognition. In B. K. Hofer & P. R. Pintrich (Eds.), *Personal epistemology: The psychology of beliefs about knowledge and knowing*. Mahway: Lawrence Erlbaum, Publisher.
Klieme, E., Hermann, A., Blum, W., et al. (Eds.) (2004). *The development of national educational standards—An expertise*. Berlin: Bundesministerium für Bildung und Forschung/Federal Ministry of Education and Research (BMBF).
Kline, M. (1980). *Mathematics: The loss of certainty*. Oxford: Oxford University Press.
Krausz, M. (Ed.). (2010). *Relativism. A contemporary anthology*. New York: Columbia University Press.
Lai, E. R. (2011). *Critical thinking: A literature review*. Retrieved from www.pearsonassessments.com/hai/images/tmrs/criticalthinkingreviewfinal.pdf.
Muis, K. R. (2004). Personal epistemology and mathematics: A critical review and synthesis of research. *Review of Educational Research Fall 2004, 74*(3), 317–377.
OECD (2010). *PISA 2009 Assessment framework*. Retrieved from www.oecd.org/pisa/pisaproducts/44455820.pdf.
Rott, B., Leuders, T., & Stahl, E. (2014). "Is mathematical knowledge certain?—Are you sure?" An interview study to investigate epistemic beliefs. *mathematica didactica, 37*, 118–132.
Rott, B., Leuders, T., & Stahl, E. (2015a). Epistemological judgments in mathematics: An interview study regarding the certainty of mathematical knowledge. In C. Bernack-Schüler, R. Erens, A. Eichler, & T. Leuders (Eds.), *Views and beliefs in mathematics education: Proceedings of the MAVI 2013 Conference* (pp. 227–238). Berlin: Springer Spektrum.

Rott, B., Leuders, T., & Stahl, E. (2015b). Assessment of mathematical competencies and epistemic cognition of pre-service teachers. *Zeitschrift für Psychologie, 223*(1), 39–46.

Rott, B., Groß Ophoff, J., & Leuders, T. (2017). Erfassung der konnotativen Überzeugungen von Lehramtsstudierenden zur Mathematik als Wissenschaft und als Schulfach. In Institut für Mathematik der Universität Potsdam (Eds.), *Beiträge zum Mathematikunterricht 2017* (pp. 1101–1104). Münster: WTM.

Schoenfeld, A. H. (1983). Beyond the purely cognitive: Belief systems, social cognitions, and metacognitions as driving focuses in intellectual performance. *Cognitive Science, 7,* 329–363.

Schoenfeld, A. H. (1985). *Mathematical problem solving*. Orlando: Academic Press.

Schoenfeld, A. H. (1989). Explorations of students' mathematical beliefs and behavior. *Journal for research in mathematics education, 20*(4), 338–355.

Schoenfeld, A. H. (2003). How can we examine the connections between teachers' world views and their educational practices? *Issues in Education, 8*(22), 217–227.

Shavelson, R. J. (2010). On the measurement of competency. *Empirical research in vocational education and training, 2*(1), 41–63.

Stahl, E. (2011). The generative nature of epistemological judgments: focusing on interactions instead of elements to understand the relationship between epistemological beliefs and cognitive flexibility (Chapter 3). In J. Elen, E. Stahl, R. Bromme, & G. Clarebout (Eds.), *Links between beliefs and cognitive flexibility—Lessons learned* (pp. 37–60). Dordrecht: Springer.

Stahl, E., & Bromme, R. (2007). The CAEB: An instrument for measuring connotative aspects of epistemological beliefs. *Learning and Instruction, 17,* 773–785.

Staub, F. C., & Stern, E. (2002). The nature of teachers' pedagogical content beliefs matters for students' achievement gains: Quasi-experimental evidence from elementary mathematics. *Journal of Educational Psychology, 94*(2), 344–355.

Stanovich, K. E., & Stanovich, P. J. (2010). A framework for critical thinking, rational thinking, and intelligence. In D. Preiss & R. J. Sternberg (Eds.), *Innovations in educational psychology: Perspectives on learning, teaching and human development* (pp. 195–237). New York: Springer.

Trautwein, U., & Lüdtke, O. (2004). Aspekte von Wissenschaftspropädeutik und Studierfähigkeit. In O. Köller, R. Watermann, U. Trautwein, & O. Lüdtke (Eds.), *Wege zur Hochschulreife in Baden-Württemberg. TOSCA – Eine Untersuchung an allgemeinbildenden und beruflichen Gymnasien* (pp. 327–366). Opladen: Leske+Budrich.

Weber, K., Inglis, M., & Meija-Ramos, J. P. (2014). How mathematicians obtain conviction: Implications for mathematics instruction and research on epistemic cognition. *Educational Psychologist, 49*(1), 36–58.

Weinert, F. E. (2001). Concept of competence: A conceptual clarification. In D. S. Rychen & L. H. Salganik (Eds.), *Defining and selecting key competencies* (pp. 45–65). Göttingen: Hogrefe.

# Epistemological Beliefs and Critical Thinking in Pre-Service Teacher Education

# 5

In Chap. 4, the first quantitative study to assess pre-service teachers' epistemological beliefs regarding the certainty of mathematical knowledge is described. In this chapter, a second, larger quantitative study is described in which the results from the previous study are tested and further elaborated and confirmed.

This chapter is based on a (peer-reviewed) journal article (Rott and Leuders 2017).[1] Because of space limits, some information regarding the German school system had to be omitted from the original article. This information is added in this chapter to specify the different strands of teacher education in Germany.

## 5.1 Background

Teacher education addresses a wide spectrum of knowledge, skills, and beliefs, which are currently regarded as aspects of the broader construct of teacher competency (Shulman 1986; Krauss et al. 2008; Blömeke et al. 2008). When focusing on those competencies that are specific for the teachers' teaching subject, various models of subject matter knowledge and pedagogical content knowledge are discussed (e.g., for mathematics Hill et al. 2004; Depaepe et al. 2013). The fact that knowledge and beliefs have an impact on teaching quality and learning outcome (Baumert et al. 2010; Voss et al. 2013) is widely acknowledged. However, the evidence on how both aspects are connected and how they develop during teacher education still mainly draws on the evidence from case studies and is in need for

[1] The article was published in the journal *JERO—Journal for Educational Research Online* (ISSN 1866-6671) by Waxmann. In their term of use, Waxmann allows the authors "[a]fter the expiration of one year after the publication of the article […] to reproduce or publish the article with reference to the primary publication in the journal JERO".



B. Rott, *Epistemological Beliefs and Critical Thinking in Mathematics*, Freiburger Empirische Forschung in der Mathematikdidaktik,
https://doi.org/10.1007/978-3-658-33539-7_5

the development of systematic measurement approaches (Blömeke et al., 2013). With respect to *mathematical knowledge*, existing instruments mostly strive to measure the amount of knowledge that has been accumulated during university curriculum (Döhrmann et al. 2014; Krauss et al. 2008). With respect to mathematical *beliefs*, one can find a broad spectrum of conceptualizations and instruments, which produce partly contradictory findings (see Sect. 5.1.1 in the theoretical background). Studies that report on development of knowledge and beliefs during teacher education typically draw on cross-sectional data and investigate the two aspects independently (Voss et al. 2013; Krauss et al. 2008; Kleickmann et al. 2013).

In our approach, we wish to contribute to the efforts to measure the development of mathematical beliefs and knowledge during the first years of university study. We are emphasizing two aspects: Firstly, we assess *both* aspects, beliefs and knowledge, simultaneously so that we can answer questions with respect to the connection between the developments in both areas. Secondly, we assess both aspects in a specific manner: Our focus is not the accumulation of subject matter knowledge, but the growing ability to flexibly solve mathematical problems. Also, we assess mathematical beliefs not by an approach that considers certain beliefs toward mathematics as naïve and others as sophisticated, but we rather differentiate between the belief position (sometimes called "belief orientation": whether mathematical knowledge is considered to be rather certain or uncertain) and the belief justification (how thoroughly the beliefs are supported by reasons and arguments).

For both purposes, we developed instruments that will be described in the following sections. We applied the instruments in several populations of students from different systems of mathematics teacher education. The intention was to test whether the findings can be interpreted consistently with the goals and the structure of the different systems and whether one can find overall structures connecting knowledge and beliefs at different points of teacher education.

Since the aim of the study is not an exhaustive definition of teacher competency or the test of a broad model, as do many large-scale-studies, we concentrate on certain dimensions, which we consider central and relevant for a better understanding of the competence structure and development in teacher education.

## 5.1.1 Theoretical Background

### 5.1.1.1 Epistemological Beliefs

With respect to the belief aspect, we focus on epistemological beliefs, which are beliefs concerning the nature of knowledge and knowing. Epistemological beliefs are a topic of psychology and (mathematics) education that has received a growing interest among researchers in recent years. The impact of epistemological beliefs on gaining and processing knowledge as well as on teaching and learning in general are widely recognized (Buehl and Alexander 2006; Hofer and Pintrich 1997). Furthermore, students' epistemological beliefs are affected by their teachers' epistemological beliefs and their teaching style (e.g., Brownlee and Berthelsen 2008).

Most researchers agree that there are several dimensions of those beliefs (cf. Hofer and Pintrich 1997) and that learners can pass several stages of development in each of those dimensions during their school and university education. Such a development is often considered to start with fixed, absolutistic beliefs and can reach flexible, cross-linked, evaluativistic beliefs. Hofer and Pintrich (1997) present an extensive review of developmental models of epistemological beliefs, stating that most researchers agree on a hierarchical sequence of stages that describes the development of such beliefs. However, there is no consensus on such stages or even on the number of stages.

However, the *theoretical* foundation as well as the empirical research on epistemological beliefs is heterogeneous (Bromme et al. 2008). Recently, beliefs are recognized as rather context-specific than general, which should be taken into account in a more systematic way (Hofer 2000). Furthermore, recent studies suggest that it is not sufficient to solely capture the general position of beliefs (e.g., "mathematical knowledge is certain vs. uncertain") but also the context in which statements on beliefs arise (Greene and Yu 2014).

In addition to these theoretical issues, there are *methodological* issues regarding the instruments used by a majority of the studies in both mathematics education and psychology as those instruments mainly consist of closed question formats (Duell and Schommer-Aikins 2001). In psychological research, the most common method of measuring epistemological beliefs is the use of questionnaires that build on Schommer's (1990) questionnaire, which uses Likert scale items (cf. Hofer 2000). In mathematics education, studies, which use closed items, are also very widespread. For example, the COACTIV study (Baumert et al. 2009, pp. 63 ff.), building on Grigutsch et al. (1998), used questions with a four-point Likert scale (is not correct/is rather not correct/is rather correct/is correct). Example items are: "Mathematics is characterized by rigor, namely a rigor in definitions and formal strictness in the mathematical argumentation." or "Mathematics is a logically consistent thought structure with precisely defined terms and uniquely provable statements." In this line of research, the expressed beliefs (e.g., "mathematics is a rigorous science") which we call "belief position" is assumed to be related to the justification of the belief. For example, beliefs on the certainty of knowledge are regarded to be absolutistic and inflexible, whereas beliefs on the uncertainty of knowledge are interpreted as evaluativistic and sophisticated (e.g., Hofer and Pintrich 1997; Schommer 1998; Muis 2004).

Stahl (2011, p. 41 f.) argues that there has been little success in developing a questionnaire with strong reliability and validity. The main problem according to Stahl is the unstable factor structure of the instruments; he identifies another problematic aspect in items, which are often indirectly related to epistemological beliefs. Muis (2004) points out additional difficulties with questionnaires in their effectiveness and in their capability of measuring general as well as domain-specific epistemological beliefs.

To overcome these theoretical and methodological issues, in an extension of the previous measuring procedures, we tried a different approach to assess epistemological beliefs. We believe that at least for the domain of mathematics, the belief position is not tied to its justification in such a strong way as the psychological research suggests (see above). For example, a person might hold the position that mathematical knowledge is uncertain and he or she might have more sophisticated arguments (e.g., the possibility of errors in the review and publication process and examples of published but erroneous proofs) or less sophisticated arguments ("Every knowledge is uncertain. I do not believe in absolute truth.") to back up his or her belief position. This view of justification is in line with Stahl's (2011) theory on cognitive flexibility and Bromme et al. (2008) who describe sophisticated epistemological beliefs "as those beliefs which allow for context-sensitive judgments about knowledge claims."

In a qualitative interview study (Rott et al. 2014; see Chap. 2), we could show that both belief positions—*mathematical knowledge is certain* vs. *uncertain*—can be held with sophisticated as well as naïve and inflexible arguments. Therefore, we strived for a quantitative instrument to capture not only individuals' belief positions but also the ways their belief positions are justified to obtain a more valid picture of peoples' beliefs. Based on the interview questions, we developed a questionnaire with open-ended questions instead of Likert-scale responses which is a precursor of the instrument described below and investigated mathematics pre-service teachers with respect to their epistemological beliefs on the certainty of mathematical knowledge. (We also investigated their mathematical abilities; see the paragraph on "Mathematical critical thinking".) In a questionnaire study with

215 pre-service teachers (Rott et al. 2015), we could show that epistemological beliefs can be measured in two dimensions—belief position and belief justification—and that these dimensions are independent of each other which at least partly contradicts previous research (see above). In our research on mathematics-related epistemological beliefs, these two belief dimensions were not related. Participants judged mathematical knowledge as either certain or uncertain either inflexibly or sophisticatedly. In that specific group, the relative frequency of sophisticated answers was higher for fourth semester students than for first semester students, indicating an increase of reflection on beliefs in this pseudo-longitudinal survey.

#### 5.1.1.2 Mathematical Critical Thinking

As a second aspect of mathematics-related competency, we focus on a dimension that relates to mathematical knowledge and its flexible application. It is not the goal of our study to comprehensively measure mathematical knowledge in teacher education (such as Baumert et al. 2010; Blömeke et al. 2008; Voss et al. 2013). Instead, we want to tap on an aspect of knowledge which reflects the flexibility of students to deal with unknown mathematical situations and which is rather independent of the mere accumulation of content knowledge. We assume that this type of mathematical flexibility is more developed in the same individuals that also argue more flexibly when asked to justify their beliefs. This assumption would be in line with the concept of a "reflective mind" as introduced below.

The instrument we developed in order to assess this aspect of mathematical knowledge was inspired by certain arguments from the research on critical thinking. Facione (1990, p. 3) "understand[s] critical thinking to be purposeful, self-regulatory judgment which results in interpretation, analysis, evaluation, and inference, as well as explanation of the evidential, conceptual, methodological, criteriological, or contextual considerations upon which that judgment is based. […]." Though many different conceptualizations of critical thinking exist (e.g., in philosophy, psychology, and education) the following abilities are commonly agreed upon (cf. Lai 2011, p. 9): analyzing arguments, claims, or evidence; making inferences using inductive or deductive reasoning; judging or evaluation and making decisions; or solving problems.

For the purpose of our study it is necessary to determine a narrower focus within this broad construct, and it seems reasonable to refer to a model by Stanovich and Stanovich (2010, p. 210 ff.). They build on *dual process theory* in which cognitive activities are distinguished into a fast, automatic, emotional, subconscious ("type 1") and a slow, effortful, logical, conscious ("type 2") subset of minds (see Kahneman 2011, for details). Stanovich and Stanovich locate critical thinking

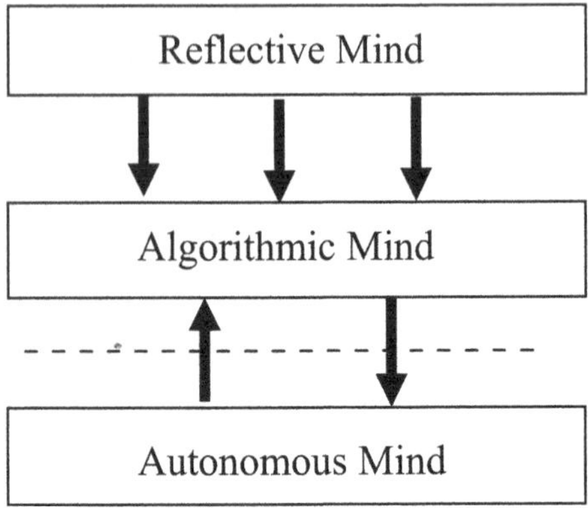


**Fig. 5.1** The tripartite model of thinking by Stanovich and Stanovich (2010, p. 210); the broken horizontal line represents the key distinction in dual process theory

within a tripartite model of thinking, an extension of dual process theory by further differentiating conscious thinking into "algorithmic" and "reflective thinking" (see Fig. 5.1). Within this model, they interpret critical thinking as a process of monitoring of problem-solving activities: Only with the reflective mind, issues of rationality come into play, assessing the efforts of the algorithmic mind. For instance, with this construct it can be explained why people with equal abilities in algorithmic thinking differ in solving complex tasks (cf. ibid., p. 212 ff.).

Within this model, an indicator of critical thinking is the ability to monitor or evaluate problem-solving processes with the reflective mind and to take more complex or less self-evident aspects of a problem into consideration.

A typical situation, which demands critical thinking to override algorithmic mathematical solutions by reflective and evaluative processes, is the paradigmatic bat-and-ball task by Kahneman and Frederick (2002): "A bat and a ball cost $ 1.10 in total. The bat costs $ 1 more than the ball. How much does the ball cost?" The spontaneous, algorithmically or even autonomously produced answer is $ 0.10. People who think critically would question this answer and realize that the ball should cost $ 0.05, whereas people who do not use critical thinking do not evaluate their first thoughts. This type of situation is typical for many mathematical problems and will be used as a model task for the construction of an instrument to measure critical mathematical thinking.

As said before, the conceptualizations of mathematical beliefs and knowledge described above are far from exhaustive with respect to mathematical competency. However, we regard them as sufficiently relevant to reflect central dimensions and to investigate their interconnection and development.

In the aforementioned study with 215 pre-service teachers (Rott et al. 2015) we additionally investigated the pre-service teachers' mathematical abilities (defined as critical thinking with respect to mathematical problem situations). We could show that the ability to think critically and the level of critical thinking does not depend on belief position but goes along with the level of justification of the epistemological judgments expressed by pre-service teachers. Also, this pseudo-longitudinal survey hinted at an increase of the ability to think critically as the fourth semester students outperformed the first semester students significantly.

## 5.2 The Study

### 5.2.1 Goals and Research Questions

The preceding study (as described in the previous section) was restricted to only 200 pre-service teachers from one university (Rott et al. 2015). Therefore, these results are considered as preliminary and will be tested with a larger number of pre-service teachers from two universities. By conducting a more comprehensive study we intend to (a) replicate the previous results, to (b) to increase their validity by applying them to different types of teacher education and to (c) test their usability as instruments for the evaluation of the outcome of tertiary education.

The research questions that guide the previous study and the one presented here are:

- Can the students' flexibility in belief justification be validly distinguished from their belief position?
- Is there a connection to the knowledge domain, more precisely their critical thinking skills operationalized as the flexibility with which students approach mathematical problems?
- Are there differences in the beliefs and the critical thinking skills between students of different numbers of semesters and different educational programs?

These questions are addressed by the following hypotheses that will be tested within the study at hand.

**Hypothesis 1** The two theoretical dimensions of epistemological beliefs—belief position (certain vs. uncertain) and belief justification (inflexible vs. sophisticated)—are empirically distinguishable.

This will be tested by means of a belief questionnaire with open-ended items in the same manner as in Rott et al. (2015). The independence of the two dimensions should be true for the whole population as well as for all relevant subgroups (i.e. low or high number of semesters and differing educational programs at the universities).

**Hypothesis 2** For the distribution of judgments regarding the certainty of mathematical knowledge, i.e. the belief positions (*certain* vs. *uncertain*), we do not expect significant differences between the relevant subgroups (i.e. low or high number of semesters and educational programs at the universities).

However, we do expect significant differences between the relevant subgroups for the distribution of the participants' belief justification (*inflexible* vs. *sophisticated*):

**Hypothesis 3** Students with a higher number of semesters argue in a more sophisticated way than students with a lower number of semesters. We also expect students with more mathematics-related content in their university education (upper secondary teachers) to argue more sophisticatedly than students with less mathematics-related content (primary and lower secondary teachers).

The flexible application of mathematical knowledge had to be measured as independently as possible of the knowledge of certain mathematical concepts learned in specific mathematics courses. To reflect this, we conceptualized mathematical abilities as critical thinking during problem solving. This ability should increase during university education, with a steeper increase in students in strands of teacher education that require more mathematics courses. Regarding the scores in the test on mathematical critical thinking, we anticipate significant differences between the subgroups:

**Hypothesis 4** Students with a higher number of semesters score higher than students with a lower number of semesters. Students with more mathematics-related content in their university education (upper secondary teachers) score higher than students with less mathematics-related content (primary and lower secondary teachers).

The assumed independence of the belief position from the justification with which beliefs are backed-up is reflected in the following hypothesis:

**Hypothesis 5** Regarding the relation of epistemological beliefs and critical thinking, there are no significant differences of critical thinking scores between the relevant subgroups sorted by their belief position (*certain* vs. *uncertain*).

The belief and the knowledge dimensions of our test both reflect certain kinds of sophistication and reflection as described in the tripartite model. Although we do not propose a model of common cognitive processes, we assume that there is a substantial correlation between these dimensions.

**Hypothesis 6** Regarding the relation of epistemological beliefs and critical thinking, there are significant differences of critical thinking scores between the relevant subgroups sorted by their belief justification (*inflexible* vs. *sophisticated*): We expect students that argue sophisticatedly in the beliefs questionnaire to score higher in the critical thinking test than students that argue inflexibly.

## 5.2.2 Methodology

### 5.2.2.1 Participants

The German school system differs within the 16 federal states of Germany, but general trends can be described as follows (for a more detailed look at this system, see König and Blömeke 2013): primary school (Grundschule) covers the grades 1–4; the secondary school is divided into three different types of schools—Hauptschule, Realschule, and Gymnasium—covering the grades 5–9, 5–10, or 5–12, respectively. The type of secondary school a student attends depends on his/her general achievement (Hauptschule for the lowest achievers) and educational goals. Attending grades 11 and 12 (upper secondary) at a Gymnasium is the usual way to gain university entrance certification.

Becoming a teacher in Germany requires an education at a university and this education depends on the type of school the future teacher aspires to teach at. The education to become a teacher at primary schools (Grundschule, grade 1–4) or lower secondary schools (Hauptschule and Realschule, grade 5–10) usually takes eight to ten semesters and comprises basic level mathematics courses (arithmetic, algebra, geometry). Becoming a teacher at upper secondary schools (Gymnasium, grade 5–13) usually takes ten semesters and requires higher level mathematics courses (analysis, linear algebra) but less educational courses compared to primary and lower secondary schools.

We present the results of a study with $n$ = 463 pre-service teachers from two universities (University of Education, Freiburg, $n$ = 277 and University of Duisburg-Essen, $n$ = 186). The students that aspire to teach at primary schools ($n$ = 198) are all enrolled at the University of Education in Freiburg; the students for upper secondary schools ($n$ = 105) are all enrolled at the University of Duisburg-Essen; the students for lower secondary schools ($n$ = 160) are enrolled either at the university in Freiburg ($n$ = 79) or in Essen ($n$ = 81). All these students participated voluntarily in this study within the first week of the 2014/15 winter term. They were contacted within lectures and were asked to complete the assessment within the lecture time.

### 5.2.3 Instruments

In this study, we identify (a) pre-service mathematics teachers' epistemological beliefs, and (b) their mathematical critical thinking and examine relationships between both constructs. Based on qualitative and quantitative preliminary studies (Rott et al. 2014, 2015), according tests and questionnaires with closed and open items have been developed.

#### 5.2.3.1 Epistemological Beliefs and Their Degree of Justification

In the study at hand, we focus on denotative beliefs, i.e. explicitly stated and reflected beliefs instead of connotative beliefs, which are affective and associative (cf. Stahl and Bromme 2007). To gain access to denotative epistemological beliefs and to the justification of the students' judgments, we constructed a questionnaire based on a preliminary interview study (Rott et al. 2014). The interviewees' answers revealed a broad spectrum of possible responses that helped us to construct according items for a quantitative test. In the resulting questionnaire, we use open-ended questions and prompts (see Table 5.1) explaining controversial points of view towards mathematics as a scientific discipline to acquire a vivid account of our participants' epistemological judgments and according arguments.

We developed a coding manual for the belief position (mathematical knowledge as *certain* vs. *uncertain*) and the level of belief justification (as *inflexible* vs. *sophisticated*). Each student's response to the four questions (a–d) in the questionnaire has been considered as one text and rated individually by two raters. These texts were analyzed with respect to the belief position and the arguments used to back up this position. A text was rated as "sophisticated" when it included arguments that reflected epistemological aspects related to the certainty of mathematical knowledge (e.g., referring to mathematical axioms, rigorous proofs,

**Table 5.1** Prompts and questions in the questionnaire regarding the certainty of mathematical knowledge

| Mathematical Knowledge is Certain | Mathematical Knowledge is Uncertain |
|---|---|
| "In mathematics, knowledge is valid forever. A theorem is never incorrect. In contrast to all other sciences, knowledge is accumulated in mathematics. [...]<br>It is impossible, that a theorem that was proven correctly will be wrong from a future point of view. Each theorem is for eternity." (Albrecht Beutelspacher) [2001, p. 235; translated by the author] | "The issue is [...] whether mathematicians can always be absolutely confident of the truth of certain complex mathematical results [...].<br>With regard to some very complex issues, truth in mathematics is that for which the vast majority of the community believes it has compelling arguments. And such truth may be fallible.<br>Serious mistakes are relatively rare, of course."<br>(Alan H. Schoenfeld) [1994, p. 58 f.] |

(a) Which of the two positions regarding the certainty of mathematical knowledge can you identify yourself with?
(b) Please, give reasons for your judgment regarding the certainty/uncertainty of mathematical knowledge
(c) Did you yourself make experiences that support one position or the other?
(d) Compare the certainty of mathematical knowledge to that of knowledge from other domains. For example, is mathematical knowledge more or less certain than knowledge from physics, language sciences, or educational sciences?

or peer-review) instead of just referring to personal opinions or knockout arguments (without an explanation for the validity of the argument). The length of an answer was not a criterion for this decision. See Table 5.2 for examples of the four possible outcomes of the rating of students' responses.

Table 5.3 shows the resulting interrater reliability; minor discrepancies occurred only when students wrote very short answers. Overall, the reliability score shows very high agreement (Cohen's $\kappa = 0.865$). After calculating the interrater reliability, the differing codes have been rated consensually by the two raters. The consensual ratings can be seen in Table 5.5.

#### 5.2.3.2 Measuring Mathematical Critical Thinking

In order to measure mathematical critical thinking as described above, more than 20 items similar to and including the bat-and-ball task have been constructed or adapted. A second example is the following task: "In a gamble, a regular six-sided die with four green faces and two red faces is rolled 20 times. You win € 25 if a certain sequence of results is shown. Which sequence would you bet on? Choose

**Table 5.2** Examples for the students' responses in the open-ended belief questionnaire

| | Inflexible | Sophisticated |
|---|---|---|
| Certain | "Mathematical knowledge is certain. Mathematical objects like numbers will never change." (SBLP-25) | "Mathematical knowledge is certain, because it is not based on observations and on theses based on these observations. Instead, mathematical knowledge is based on conventions (axioms) and resulting theorems. There may be differences regarding these conventions but not the conclusions." (ESEB-16) |
| Uncertain | "Mathematical knowledge is uncertain, because arguments can be rebutted." (LAAI-21) | "Mathematical knowledge is uncertain. A proof is only valid, because a majority of humans considers it and the according arguments as valid. Without this approval, it would not be valid anymore. Therefore, the certainty of a proof depends on the judgment of humans and this is not safe." (BJAT-16) |

**Table 5.3** Interrater agreement coding "Certainty of Mathematical Knowledge" (between R1 and R2; R1 being the author of this book)

| | | R1 | | | | Sum |
|---|---|---|---|---|---|---|
| | | Certain & inflexible | Certain & sophisticated | Uncertain & inflexible | Uncertain & sophisticated | |
| R2 | Certain & inflexible | 203 | 0 | 25 | 1 | 229 |
| | Certain & sophisticated | 0 | 17 | 0 | 0 | 17 |
| | Uncertain & inflexible | 0 | 0 | 192 | 5 | 197 |
| | Uncertain & sophisticated | 0 | 0 | 25 | 15 | 20 |
| | Sum | 203 | 17 | 222 | 21 | 463 |

$P_{obs} = 0.922$, $P_{exp} = 0.424$, Cohen's $\kappa = 0.87$

one: (a) RGRRR (b) GRGRRR (c) GRRRRR." Most students choose option (b) because it contains more instances of G than the other options. Those students do not realize that option (a) is entirely included in option (b) and, therefore, more likely.

All items were related to mathematical situations and demanded only knowledge from lower secondary education. In four preliminary studies—of which three were quantitative studies and one was a qualitative study using task-based interviews—(Rott and Leuders 2016), it was investigated whether the items measure computational skills or the willingness to engage in a critical reflection of apparently obvious solutions. Items that did not trigger critical thinking according to the model by Stanovich and Stanovich were discarded. After validation and because we wanted to restrict the time for this test to 20 min, the final test consisted of 11 items (for further details see Rott et al. 2015).

All items were rated dichotomously and we used a Rasch model to transform our participants' test scores into values on a one-dimensional competency scale (software RUMM 2030 by Andrich et al. 2009). After eliminating two items because of underdiscrimination (fit residual > 2.5) in connection with floor and ceiling effects, respectively, for each item the model showed good fit residuals (all values between −2.5 and 2.5) and no significant differences between the observed overall performance of each trait group and its expected performance (overall-$\chi^2 = 36.2$; df = 27; $p = 0.11$).

### 5.2.4 Results

Firstly, we present the distribution of the pre-service teachers' responses regarding the questionnaire on epistemological beliefs. Secondly, we evaluate our participants' scores on the test of mathematical critical thinking. Thirdly, we investigate possible relations between both aspects of mathematics-related research competency.

#### 5.2.4.1 Denotative Beliefs: Position and Justification

The distribution of students that filled out the belief questionnaire is presented in Table 5.4. The data is sorted by the four possible outcomes of the two belief dimensions (belief position combined with the degree of justification). Additionally, the students' distribution has been sorted by their number of semesters (into novice students with three or less semesters and into advanced students with four or more semesters respectively, this also happens to be a median split) as well as

**Table 5.4** Distribution of the students' denotative epistemological beliefs and their degree or justification

| | Certain & inflexible | Certain & sophisticated | Uncertain & inflexible | Uncertain & sophisticated | Sum |
|---|---|---|---|---|---|
| Semester ≤ 3 | 117 (47.4 %) | 6 (2.4 %) | 114 (46.2 %) | 10 (4.0 %) | 247 (100 %) |
| Semester ≥ 4 | 86 (39.8 %) | 11 (5.1 %) | 108 (50.0 %) | 11 (5.1 %) | 216 (100 %) |
| Primary | 84 (42.4 %) | 5 (2.5 %) | 97 (49.0 %) | 12 (6.1 %) | 198 (100 %) |
| Lower Secondary | 67 (41.9 %) | 4 (2.5 %) | 85 (53.1 %) | 4 (2.5 %) | 160 (100 %) |
| Upper Secondary | 52 (49.5 %) | 8 (7.6 %) | 40 (38.1 %) | 5 (4.8 %) | 105 (100 %) |
| All students combined | 203 (43.8 %) | 17 (3.7 %) | 222 (47.9 %) | 21 (4.5 %) | 463 (100 %) |

by their aspired teaching profession (primary, lower secondary, or upper secondary school). The last row shows the total number of students in each of the four belief categories. The low percentage of students arguing sophisticatedly (8.2 %) is in line with the respective percentage of students (12.9 %) in the preliminary quantitative study (Rott et al. 2015).

Chi-square tests were used to address the question whether the two theoretically claimed dimensions (belief position and according justification, cf. Hypothesis 1) are independent. Table 5.5 presents the set-up of the data for the chi-square test for all students which are the same numbers as in the last row of Table 5.4 (Yates chi-square test, corrected for continuity: $\chi^2 = 0.04$, df = 1, p = 0.84).

**Table 5.5** Comparison of both dimensions of the belief questionnaire for all students; the numbers in parentheses indicate expected frequency under the assumption of statistical independence

| | Inflexible | Sophisticated | Sum |
|---|---|---|---|
| Certain | 203 (201.9) | 17 (18.1) | **220** |
| Uncertain | 222 (223.1) | 21 (19.9) | **243** |
| Sum | **425** | **38** | **463** |
| | $\chi^2 = 0.04$ | df = 1 | $p = 0.842$ |

The chi-square test has also been repeated for all sub-groups to check for possible group specific dependencies (Semester ≤ 3: $\chi^2 = 0.58$, p = 0.446; Semester ≥ 4: $\chi^2 = 0.08$, p = 0.777; Primary schools: $\chi^2 = 1.19$, p = 0.275; Lower secondary schools: $\chi^2 = 0.001$, p = 0.975 (numbers too small for exact results); Upper secondary schools: $\chi^2 = 0.002$, p = 0.964; all df = 1). These results confirm our hypothesis 1 that both dimensions are statistically independent. Therefore, in Table 5.6 we sort the numbers of students by those two dimensions separately.

A *first question* regarding these data concerns the pre-service teachers' judgments: Do they regard mathematical knowledge as certain or uncertain and does this change with the number of semesters (cf. Hypothesis 2)? The ratio of pre-service teachers judging "certain" in semester 3 or less is 123:124 (49.8 % certain) compared to 97:119 (44.9 %) in semester 4 or higher; there is no significant deviation from the null-hypothesis ("no effect") (Yates chi-square test, corrected for continuity: $\chi^2 = 0.92$, df = 1, $p = 0.338$).

This question can be repeated for each sub-group. For pre-service teachers of primary schools, there is no significant difference in the judgment of mathematical knowledge as "certain" in semester 3 or less (42:43, 49.4 % certain) compared to students in semester 4 or higher (47:66, 41.6 % certain) ($\chi^2 = 0.90$, df = 1, $p = 0.343$). There is also no significant difference for pre-service teachers of lower secondary schools in semester 3 or less (51:69, 42.5 % certain) compared to students in semester 4 or higher (20:20, 50.0 % certain) ($\chi^2 = 0.41$, df = 1, $p = 0.522$). There is, however, a significant difference for pre-service teachers of upper secondary schools in semester 3 or less (30:12, 71.4 % certain) compared to

**Table 5.6** Distribution of the students' denotative epistemological beliefs and their degree or justification; sorted by the two dimensions separately

| | Certain | Uncertain | Inflexible | Sophisticated | Total |
|---|---|---|---|---|---|
| Semester ≤ 3 | 123 (49.8 %) | 124 (50.2 %) | 231 (93.5 %) | 16 (6.5 %) | 247 (100 %) |
| Semester ≥ 4 | 97 (44.9 %) | 119 (55.1 %) | 194 (89.8 %) | 22 (10.2 %) | 216 (100 %) |
| Primary | 89 (44.9 %) | 109 (55.1 %) | 181 (91.4 %) | 17 (8.6 %) | 198 (100 %) |
| Lower secondary | 71 (44.4 %) | 89 (55.6 %) | 152 (95.0 %) | 8 (5.0 %) | 160 (100 %) |
| Upper secondary | 60 (57.1 %) | 45 (42.9 %) | 92 (87.6 %) | 13 (12.4 %) | 105 (100 %) |
| All students combined | 220 (7.5 %) | 243 (2.5 %) | 425 (1.8 %) | 38 (8.2 %) | 463 (100 %) |

students in semester 4 or higher (30:33, 47.6 % certain) ($\chi^2 = 4.90$, df = 1, $p =$ 0.027). Except for the pre-service teachers that aspire to teach at upper secondary schools, there are no differences within the subgroups, which is in accordance with Hypothesis 2.

A *second question* concerns the participants' justification: Do they argue inflexibly or sophisticatedly and does the belief justification differ between the relevant subgroups (number of semesters and educational program, cf. Hypothesis 3)? The number of sophisticated reasoning is higher for students with a higher number of semesters than for students with a lower number of semesters, which is 194:22 (inflexible:sophisticated, i.e. 10.2 % sophisticated) (semester $\geq$ 4) in comparison to 231:16 (6.5 %) (semester $\leq$ 3).

A closer inspection of the sub-groups shows that this effect is not visible in primary students: A larger percentage of students of semester 3 or less (76:9, 10.6 % sophisticated) argues sophisticatedly compared to students of semester 4 or higher (105:8, 7.1 % sophisticated). The effect is visible, though, for the other two sub-groups: Pre-service teachers of lower secondary schools in semester 3 or less (117:3, 2.5 % sophisticated) show less sophisticated statements compared to students in semester 4 or higher (35:5, 12.5 % sophisticated). Pre-service teachers of upper secondary schools in semester 3 or less (38:4, 9.5 % sophisticated) also show less sophisticated belief justifications compared to students in semester 4 or higher (54:9, 14.3 % sophisticated). Except for the primary school pre-service teachers, the slow shift towards more sophisticated belief justification is in accordance with Hypothesis 3.

A closer look at the students in the different educational programs confirms the hypothesis, that a higher percentage of students that aspire to become secondary teachers argue sophisticatedly than students for lower secondary and primary schools (12.5 % compared to 5.0 % and 8.6 %, respectively).

#### 5.2.4.2 Critical Thinking

The Rasch model of the students' critical thinking ability provides metrical latent variables ranging from −2.83 to 2.81 with low values indicating a low ability. Table 5.7 presents the means of these values that have been sorted by aspired type of school and by the number of semesters (cf. Rott et al. 2015).

A t-test was used to compare the two groups that aspire to teach at lower secondary schools (Freiburg: −0.222 (0.864), $n = 79$; Essen: −0.255 (1.047), $n = 81$). The test showed no significant differences between these two groups ($t =$ 0.22, df = 158, $p_{\text{two-tailed}} = 0.826$), so that we also combined these two groups in our analyses when necessary.

**Table 5.7** Means (and standard deviations) of Mathematical Critical Thinking, sorted by aspired teaching position (crossing universities) and the number of semesters

| | Semester $\leq 3$ | Semester $\geq 4$ | Total |
|---|---|---|---|
| Primary | 0.031 (0.872) | −0.369 (0.915) | −0.197 (0.916) |
| | $n = 85$ | $n = 113$ | $n = 198$ |
| Lower secondary | −0.235 (0.868) | −0.252 (1.202) | −0.239 (0.958) |
| | $n = 120$ | $n = 40$ | $n = 160$ |
| Upper secondary | −0.117 (0.907) | 0.354 (0.995) | 0.166 (0.984) |
| | $n = 42$ | $n = 63$ | $n = 105$ |
| Total | −0.123 (0.881) | −0.136 (1.041) | −0.129 (0.958) |
| | $n = 247$ | $n = 216$ | $n = 463$ |

A question that can be addressed to these data is whether critical thinking scores are dependent on the semester and the aspired teaching position (cf. Hypothesis 4). Surprisingly, there is no significant advantage in the critical thinking scores for students with a higher number of semesters, which could be assumed (and was visible in the preliminary study). Actually, the three study programs show very different results regarding the development of critical thinking within this pseudo-longitudinal survey: Only in the group of students aspiring to teach at upper secondary schools, the more experienced students show significantly higher critical thinking scores (t-test: $t$ = 2.46, df = 103, $p_{2\text{-sided}}$ = 0,016; Cohen's $d$ = 0.48). For lower secondary schools, there is no significant difference ($t$ = 0.1, df = 158, p2-sided = 0,921), and for primary schools, there is even a decline ($t$ = 3.1, df = 196, $p_{2\text{-sided}}$ = 0,002; Cohen's $d$ = 0.44). Therefore, it is not recommendable to interpret the results of a two-way ANOVA that has been used to investigate whether there are differences between students of semester three or less compared to students of semester four or greater, and whether there are differences between students of the different study programs.

The part of Hypothesis 4 regarding higher critical thinking scores for students with a higher number of semesters can therefore not be confirmed. The critical thinking scores regarding the aspired teaching position, however, indicate that students show higher scores in the mathematical critical thinking test, if they have the more profound mathematics education (upper secondary vs. lower secondary and primary education).

#### 5.2.4.3 The Relation of Epistemological Beliefs and Critical Thinking

Table 5.8 presents mean values of critical thinking scores in relation to the students' answers on the epistemological belief questionnaire; the numbers of students are identical to those in Table 5.6.

A *first question* that can be addressed to these data is whether critical thinking scores depend on the belief position ("certain" vs. "uncertain") (cf. Hypothesis 5). A two-way ANOVA has been used to answer this question: There are no significant differences between students regarding mathematical knowledge as "certain" compared to students regarding it as "uncertain" ($F = 0.159$; $p = 0.690$). There are significant differences between students of the different study programs (see above) ($F = 6.200$; $p = 0.002$); a Tukey Post Hoc reveals that the differences between the study programs are significant between primary and upper secondary ($p = 0.002$) as well as between lower and upper secondary ($p = 0.004$), but not between primary and lower secondary ($p = 0.907$) education. Also, there is a

**Table 5.8** Means (and standard deviations) of Mathematical Critical Thinking, sorted by the dimensions of the belief questionnaire

| | Certain | Uncertain | Inflexible | Sophisticated | Total |
|---|---|---|---|---|---|
| Semester ≤ 3 | −0.125 (0.866) | −0.121 (0.898) | −0.163 (0.858) | 0.451 (1.029) | −0.123 (0.881) |
| | $n = 123$ | $n = 124$ | $n = 231$ | $n = 16$ | $n = 247$ |
| Semester ≥ 4 | −0.076 (1.089) | −0.185 (1.003) | −0.195 (1.026) | 0.384 (1.052) | −0.136 (1.041) |
| | $n = 97$ | $n = 119$ | $n = 194$ | $n = 22$ | $n = 216$ |
| Primary | −0.129 (0.921) | −0.253 (0.913) | −0.239 (0.889) | 0.245 (1.103) | −0.197 (0.916) |
| | $n = 89$ | $n = 109$ | $n = 181$ | $n = 17$ | $n = 198$ |
| Lower secondary | −0.397 (1.036) | −0.113 (0.877) | −0.271 (0.936) | 0.370 (1.235) | −0.239 (0.958) |
| | $n = 71$ | $n = 89$ | $n = 152$ | $n = 8$ | $n = 160$ |
| Upper secondary | 0.282 (0.832) | 0.011 (1.148) | 0.096 (0.991) | 0.656 (0.803) | 0.166 (0.984) |
| | $n = 60$ | $n = 45$ | $n = 92$ | $n = 13$ | $n = 105$ |
| All students combined | −0.104 (0.969) | −0.153 (0.949) | −0.178 (0.938) | 0.412 (1.029) | −0.129 (0.958) |
| | $n = 220$ | $n = 243$ | $n = 425$ | $n = 38$ | $n = 463$ |

significant interaction effect ($F = 3.263$; $p = 0.039$). The results regarding the study programs and the interaction effect have to be interpreted carefully because of the disordinal interaction between study program and number of semesters (see above). Nonetheless, these results confirm our Hypothesis 5 that critical thinking scores are not related to the belief position (*certain* vs. *uncertain*).

The *second question* in this context addresses the correlation of critical thinking scores with the degree of belief justification (cf. Hypothesis 6). Another ANOVA shows that students that argue sophisticatedly score significantly better than students that argue in an inflexible way ($F = 11.386$; $p = 0.001$; $d = 0.62$; observed statistical power[2] of this effect is 96.1 %). A closer look at the data shows that the significant differences for the belief justification are true for all three subgroups of aspired teaching positions (see also Table 5.8). Thus, the results confirm our Hypothesis 6 that there is a marked connection between justification of beliefs and the ability of thinking critically when solving tasks.

### 5.2.5 Discussion

#### 5.2.5.1 Discussion of the Hypotheses

In the study at hand, most of the hypotheses that were drawn from the preliminary study (Rott et al. 2015) could be confirmed.

Hypothesis 1, the assumed independence of belief position and justification, has been *confirmed* for the whole test population as well as for each sub-group (pre-service teachers with a lower and higher number of semesters as well as for each of the three aspired school types). This result is particularly important for research on epistemological beliefs since within closed questionnaire surveys the belief position is often used to determine its level of justification. For example, Hofer and Pintrich (1997, p. 119 f.) assume that the belief "absolute truth exists with certainty" is valid only for "lower levels" of belief justification (i.e. it is less sophisticated). Our study shows that the belief position of certainty can be held and supported also with sophisticated arguments—at least in the area under investigation. Further areas remain to be investigated to corroborate the assumption of independence.

Hypothesis 2 stated that there are no significant differences regarding belief position within the sub-groups (low or high number of semesters and educational programs at the universities). This hypothesis was *confirmed* except for

[2] Post hoc statistical power, calculated online with a program provided by DSS at www.dssresearch.com/KnowledgeCenter/ToolkitCalculators.

the group of upper secondary pre-service teachers. We can only speculate on possible reasons of the change within this group. It might be due to the strong mathematics-related content and experiences of flawed proofs of these students' education that favors a shift to uncertainty beliefs. However, we conducted no genuine longitudinal survey and are unable to capture actual changes in students' beliefs.

Hypothesis 3 predicted differences within the sub-groups regarding the belief justification. This hypothesis was *confirmed* except for the group of primary pre-service teachers. Older students should be better able to argue sophisticatedly against a richer backdrop of knowledge and experience, which has become evident in the data. Also, students with a richer mathematical background (trainees for upper secondary schools) showed a higher percentage of sophisticated justification. We do not know, however, what happened in the sample of pre-service teachers for primary schools.

Hypothesis 4, the assumed increase in critical thinking scores between the students from lower and higher semesters could *not* be *confirmed*; but it was *confirmed* for the different educational programs. The unexpected stagnation with respect to critical thinking in some of the groups may be investigated further with respect to possible group substructures. A reason for this could be anything from a bad day that influenced test results for that specific day to lectures that sustainably affected the willingness to think critically of some participants. However, the present data does not allow for such a further investigation. One may argue that the tasks used in the critical thinking test do not require mathematics that are taught at the university level and should therefore not depend on the number of semesters.

Hypothesis 5 and Hypothesis 6, predicting no relations between belief position and mathematical competency (operationalized by our critical thinking test) but significant relations between belief justification and mathematical competency could both be *confirmed*. Picking up the discussion from Hypothesis 1, this result seems to be of high importance. Studies trying to show connections between (epistemological) beliefs and other factors of (mathematical) competencies would be well advised to collect data regarding belief justification not solely with instruments that operationalize it via belief positions. Hypothesis 5 and Hypothesis 6 also allow for hypotheses regarding the development of students' competencies during their university education for subsequent studies.

#### 5.2.5.2 Limitations

Due to our specific procedure, the study has several limitations.

Firstly, some considerations with respect to the validity of the testing procedures seem to be necessary: Within our study, we assumed the perspective of distinguishing between beliefs in a dichotomous manner. This is partly inherent in the design (participants had to decide between two opposing statements) and partly due to the evaluation methods (dichotomization of the multivariate data). Studies with experts (e.g., Mura 1993) indicate that this approach becomes invalid, when the individuals have broad experience in their subject and hold complex and differentiated views. For people that have reached a very high degree of justification, such as professional mathematicians, the method of forced choice between two belief positions (e.g., "certain" vs "uncertain") may lead to invalidities, since such people tend to answer that *both* positions can be adequate depending on the context (Stahl 2011; Gowers 2013). The interviews that were conducted in our preliminary qualitative study (Rott et al. 2014) confirm these considerations: highly sophisticated interviewees refuse to commit themselves to one belief position. However, in those interviews (ibid.) it was found that for students such a dialectic position is not yet in reach and that deciding for one or the other belief position is not triggering irritations. Therefore, we consider the methodological decision to rate the students' belief positions dichotomously as reasonable for the study at hand.

As stated above, all differences regarding the number of the pre-service teachers' semester have to be interpreted with care, as we did not conduct a longitudinal survey. Within this study, we cannot trace the development of students' epistemological beliefs in the course of their university studies. Furthermore, we cannot tell whether results in favor of students with more semesters are due to a gain in knowledge or to selection effects, i.e. low-performing students leaving the university. To answer according questions, follow-up studies have to be conducted. For a validation, it would be most desirable to use the instrument for investigating change of belief and critical thinking during specific learning environments.

Also, the strands may be confounded with the location. Such considerations can be dealt with when we extend our investigations to more than only two universities.

#### 5.2.5.3 Theoretical Considerations with Respect to Epistemological Beliefs and Critical Thinking

We do not offer a model that explains the justification of beliefs and the level of critical thinking. Possibly, we deal here with two quite different constructs (one more verbal, the other more mathematical) that grow simultaneously. However, Kuhn (1999, p. 22 f.) discusses possible relationships between the development of epistemological beliefs and critical thinking. She argues that the transition from absolutistic to evaluativistic beliefs of the knowledge structure and the ability to think critically are closely related. On the one hand, critical reflections lead to questioning beliefs and to the insight that even experts disagree about important issues. These are important steps in developing more sophisticated epistemological beliefs. On the other hand, an absolutist epistemological understanding favors easy and more direct answers on questions of truth or falsity. Kuhn concludes that individuals who confine themselves to an absolutist epistemology have a low demand for critical thinking skills and, hence, the impetus to exercise and further develop these skills is slight.

The study presented here is conducted in mathematics teacher education. However, we assume that the constructs presented here can also be found with respect to other subjects, so that it would be interesting to ask whether similar findings would be encountered e.g., in the education of science or history teachers. Still, the constructs would partly need different operationalizations (e.g., for critical thinking) or a different emphasis on epistemological aspects of the subject than those relevant for mathematics.

## References


Andrich, D., Sheridan, B. E., & Luo, G. (2009). *RUMM2030: Rasch unidimensional models for measurement*. Perth: RUMM Laboratory.

Baumert, J., Blum, W., Brunner, M., Dubberke, T., Jordan, A., Klusmann, U., et al. (2009). *Professionswissen von Lehrkräften, kognitiv aktivierender Mathematikunterricht und die Entwicklung von mathematischer Kompetenz (COACTIV): Dokumentation der Erhebungsinstrumente*. Berlin: Max-Planck-Institute for Human Development.

Baumert, J., Kunter, M., Blum, W., Brunner, M., Voss, T., Jordan, A., et al. (2010). Teachers' mathematical knowledge, cognitive activation in the classroom, and student progress. *American Educational Research Journal, 47*(1), 133–180.

Blömeke, S., Kaiser, G., & Lehmann, R. (Eds.). (2008). *Professionelle Kompetenz angehender Lehrerinnen und Lehrer. Wissen, Überzeugungen und Lerngelegenheiten deutscher Mathematikstudierender und -referendare. Erste Ergebnisse zur Wirksamkeit der Lehrerausbildung*. Münster: Waxmann.

Blömeke, S., Zlatkin-Troitschanskaia, O., Kuhn, C., & Fege, J. (2013). *Modeling and measuring competencies in higher education* (pp. 1–10). Sense Publishers.

Bromme, R., Kienhues, D., & Stahl, E. (2008). Knowledge and epistemological beliefs: An intimate but complicate relationship. In M. S. Khine (Ed.), *Knowing, knowledge and beliefs. Epistemological studies across diverse cultures* (pp. 423–441). New York: Springer.

Brownlee, J., & Berthelsen, D. (2008). Developing relational epistemology through relational pedagogy. In M. S. Khine (Ed.), *Knowing, knowledge and beliefs. Epistemological studies across diverse cultures* (pp. 405–422). New York: Springer.

Buehl, M. M., & Alexander, P. A. (2006). Examining the dual nature of epistemological beliefs. *International Journal of Educational Research, 45,* 28–42.

Depaepe, F., Verschaffel, L., & Kelchtermans, G. (2013). Pedagogical content knowledge: A systematic review of the way in which the concept has pervaded mathematics educational research. *Teaching and Teacher Education, 34,* 12–25.

Döhrmann, M., Kaiser, G., & Blömeke, S. (2014). The conceptualisation of mathematics competencies in the international teacher education study TEDS-M. In S. Blömeke, F.-J. Hsieh, G. Kaiser, & W. H. Schmidt (Eds.), *International perspectives on teacher knowledge, beliefs and opportunities to learn* (pp. 431–456). Dordrecht: Springer.

Duell, O. K., & Schommer-Aikins, M. (2001). Measures of people's beliefs about knowledge and learning. *Educational Psychology Review, 13,* 419–449.

Facione, P. A. (1990). *Critical thinking: A statement of expert consensus for purposes of educational assessment and instruction [Executive Summary "The Delphi Report"].* Millbrae: California Academic Press.

Gowers, T. (2013). Is mathematics discovered or invented? In M. Pitici (Ed.), *The best writing on mathematics* (pp. 8–20). Princeton und Oxford: Princeton University Press.

Greene, J. A., & Yu, S. B. (2014). Modeling and measuring epistemic cognition: A qualitative re-investigation. *Contemporary Educational Psychology, 39,* 12–28.

Grigutsch, S., Raatz, U., & Törner, G. (1998). Einstellungen gegenüber Mathematik bei Mathematiklehrern. *Journal für Mathematik-Didaktik, 19*(1), 3–45.

Hill, H. C., Schilling, S. G., & Ball, D. L. (2004). Developing measure of teachers' mathematics knowledge for teaching. *The Elementary School Journal, 105*(1), 11–30.

Hofer, B. K. (2000). Dimensionality and disciplinary differences in personal epistemology. *Contemporary Educational Psychology, 25,* 378–405.

Hofer, B. K., & Pintrich, P. R. (1997). The development of epistemological theories: Beliefs about knowledge and knowing and their relation to learning. *Review of Educational Research 1997, 67*(1), 88–140.

Kahneman, D., & Frederick, S. (2002). Representativeness revisited: Attribute substitution in intuitive judgment. In T. Gilovich, D. Griffin, & D. Kahneman (Eds.), *Heuristics and biases: The psychology of intuitive judgment* (pp. 49–81). New York: Cambridge University Press.

Kahneman, D. (2011). *Thinking, fast and slow.* London: Penguin Books Ltd.

Kleickmann, T., Richter, D., Kunter, M., Elsner, J., Besser, M., Krauss, S., et al. (2013). Pedagogical content knowledge and content knowledge of mathematics teachers: The role of structural differences in teacher education. *Journal of teacher education, 64,* 90–106.

König, J., & Blömeke, S. (2013). Preparing teachers of mathematics in Germany. In J. Schwille, L. Ingvarson, & R. Holdgreve-Resendez (Eds.), *TEDS-M encyclopaedia. A guide to teacher education context, structure and quality assurance in 17 countries. Findings from the IEA teacher education and development study in mathematics (TEDS-M)* (pp. 100–115). Amsterdam: IEA.

Krauss, S., Brunner, M., Kunter, M., Baumert, J., Blum, W., Neubrand, M., et al. (2008). Pedagogical content knowledge and content knowledge of secondary mathematics teachers. *Journal of Educational Psychology, 100*(3), 716–725.

Kuhn, D. (1999). A developmental model of critical thinking. *Educational Researcher, 28*(2), 16–25 + 46.

Lai, E. R. (2011). *Critical thinking: A literature review*. Upper Saddle River: Pearson Assessment. Retrieved from www.pearsonassessments.com/hai/images/tmrs/criticalthinkingreviewfinal.pdf.

Muis, K. R. (2004). Personal epistemology and mathematics: A critical review and synthesis of research. *Review of Educational Research, 74,* 317–377.

Mura, R. (1993). Images of mathematics held by university teachers of mathematical sciences. *Educational Studies in Mathematics, 25*(4), 375–385.

Rott, B., Leuders, T., & Stahl, E. (2014). "Is mathematical knowledge certain?—Are you sure?" An interview study to investigate epistemic beliefs. *mathematica didactica, 37,* 118–132.

Rott, B., Leuders, T., & Stahl, E. (2015). Assessment of mathematical competencies and epistemic cognition of pre-service teachers. *Zeitschrift für Psychologie, 223*(1), 39–46.

Rott, B., & Leuders, T. (2016). Mathematical critical thinking: The construction and validation of a test. In C. Csíkos, A. Rausch & J. Szitányi (Eds.), *Proceedings of the 40th conference of the international group for the psychology of mathematics education* (Vol. 4, pp. 139–146). Szeged: PME.

Rott, B., & Leuders, T. (2017). Mathematical competencies in higher education: Epistemological beliefs and critical thinking in different strands of pre-service teacher education. *JERO—Journal for Educational Research Online, 9*(2), 113–134.

Schommer, M. (1998). The role of adults' beliefs about knowledge in school, work, and everyday life. In M. C. Smith & T. Pourchot (Eds.), *Adult learning and development—perspectives from educational psychology* (pp. 127–143). London: Lawrence Erlbaum.

Schommer, M. (1990). Effects of beliefs about the nature of knowledge comprehension. *Journal of Educational Psychology, 82*(3), 498–504.

Shulman, L. S. (1986). Those who understand: Knowledge growth in teaching. *Educational researcher*, 4–14.

Stahl, E. (2011). The generative nature of epistemological judgments: Focusing on interactions instead of elements to understand the relationship between epistemological beliefs and cognitive flexibility. In J. Elen, E. Stahl, R. Bromme, & G. Clarebout (Eds.), *Links between beliefs and cognitive flexibility—lessons learned* (pp. 37–60). Dordrecht: Springer.

Stahl, E., & Bromme, R. (2007). The CAEB: An instrument for measuring connotative aspects of epistemological beliefs. *Learning and Instruction, 17,* 773–785.

Stanovich, K. E., & Stanovich, P. J. (2010). A framework for critical thinking, rational thinking, and intelligence. In D. Preiss & R. J. Sternberg (Eds.), *Innovations in educational psychology: Perspectives on learning, teaching and human development* (pp. 195–237). New York: Springer.

Voss, T., Kleickmann, T., Kunter, M., & Hachfeld, A. (2013). Mathematics teachers' beliefs. In M. Kunter, J. Baumert, W. Blum, U. Klusmann, S. Krauss, & M. Neubrand (Eds.), *Cognitive activation in the mathematics classroom and professional competence of teachers—results from the COACTIV project* (pp. 249–272). New York: Springer.

# Inductive and Deductive Justification of Knowledge

# 6

In this chapter, a second part of the large quantitative study that is described in Chap. 5 is presented. The same students that answered questionnaire items regarding beliefs about the *certainty of mathematical knowledge* were asked afterwards to reply to items regarding beliefs about the *justification of mathematical knowledge*. However, not all students that replied to the first group of items did so for the second group. Therefore, the number of participants is slightly lower.

Similar to the questionnaire items regarding beliefs about the certainty of *knowledge*, the items regarding beliefs about the *justification of knowledge* were developed by means of the questionnaire study that is presented in Chap. 4.

This chapter is based on a (peer-reviewed) journal article (Rott and Leuders 2016a, b).[1] The terminology was slightly adapted to better fit to the other chapters in this book.

---

[1] The article (doi https://doi.org/10.1080/10986065.2016.1219933) was published in the journal *MTL—Mathematical Thinking and Learning* (ISSN 1098-6065) by Routledge—Taylor & Francis Group. In their term of use, the publisher states: "The right to expand your article into book-length form for publication provided that acknowledgement to prior publication in the Journal is made explicit (see below). Where permission is sought to re-use an article in a book chapter or edited collection on a commercial basis a fee will be due, payable by the publisher of the new work. Where you as the author of the article have had the lead role in the new work (i.e., you are the author of the new work or the editor of the edited collection), fees will be waived."



B. Rott, *Epistemological Beliefs and Critical Thinking in Mathematics*, Freiburger Empirische Forschung in der Mathematikdidaktik, https://doi.org/10.1007/978-3-658-33539-7_6

## 6.1 Background

### 6.1.1 Introduction

Studying individuals' belief systems has a long tradition in educational research. A person's beliefs are his or her "subjective knowledge […] of a certain object or concern for which indisputable grounds may not necessarily be found in objective considerations." (Törner and Pehkonen 1999, p. 1)

In mathematics education, the research of Schoenfeld (1983) on students' beliefs on the topic of problem solving is considered a seminal work on the inter-relation of mathematical beliefs and learning processes. Furthermore, recent studies focus on beliefs of teachers because of their assumed influence on students' learning processes (cf. Philipp 2007; Staub and Stern 2002). For example, Nardi et al. (2012) showed that pedagogical and epistemological beliefs determined mathematics teachers' examinations of student solutions.

Research on beliefs from mathematics education as well as from educational psychology has put a special emphasis on *epistemological beliefs*, which are beliefs about the nature of knowledge and knowing (cf. Hofer and Pintrich 1997). Since the 1990s, these beliefs are considered as a multifaceted construct, incorporating several more or less independent beliefs or belief dimensions (Hofer and Pintrich 1997; Hofer 2000). Examples of such beliefs concern the source, the certainty, or the justification of knowledge. According to Muis (2004), many students believe that the source of mathematical knowledge is their teacher or that knowledge is justified "on a purely empirical approach […] ignor[ing] a more rational and logical approach that would have included the use of proofs in the discovery and verification process." (p. 340). Beliefs like these shape the students' approaches to learning, and influence their academic achievement.

Epistemological beliefs are considered to be an important educational goal as they are assumed to be relevant whenever an individual has to evaluate knowledge with respect to its reliability (e.g., Bromme et al. 2008). Research indicates that more sophisticated epistemological beliefs are related to more suitable learning strategies and, therefore, better learning outcomes (Schommer 1993; Hofer and Pintrich 1997; Trautwein and Lüdtke 2007; Stahl 2011). Teachers' conceptions of subject matter and their epistemological beliefs play an important role in shaping their classroom practice (Fennema and Franke 1992). Additionally, it has been shown that epistemological beliefs of teachers have an impact on their students' beliefs (Buelens et al. 2002; Tsai 1998). However, there is conflicting evidence that cannot be explained with most of the established theories about epistemological beliefs: Even though the constructs representing epistemological beliefs are considered as general and rather stable by most researchers, there is a growing body of empirical evidence that epistemological beliefs are less coherent, more domain-specific and more context-dependent than previously assumed (Buehl and Alexander 2006; de Corte et al. 2002; Pintrich 2002; Stahl and Bromme 2007).

Furthermore, there are methodological issues regarding the techniques of assessing epistemological beliefs. Especially self-report instruments such as commonly used questionnaires with rating scales seem to lack validity (Greene and Yu 2014; Muis 2004).

We try to overcome these conceptual and methodological issues by differentiating between belief orientation on the one hand and its sophistication on the other hand, thus accounting for the flexibility of personal epistemology as proposed by e.g., Elby and Hammer (2010) or Stahl (2011). To be more precise, we assume that there is a distinction to be made between someone holding a belief position and his or her explicit efforts to justify this position. Previous research has already shown this by qualitative analysis of interviews (Rott et al. 2014) and by statistical analyses of questionnaire data (Rott et al. 2015) regarding the certainty of mathematical knowledge: A student can assume mathematical knowledge either to be rather certain or uncertain independently of the sophistication with which he or she is able to support the respective judgement. Stimulated by these findings, we present two related studies (an interview study and a questionnaire study) that investigate beliefs on the justification of mathematical knowledge. We have a focus on pre-service teachers but also include teachers and teacher educators. These studies are intended to show that both dimensions—the belief position and its sophistication in the area of justification of mathematical knowledge—are rather independent of each other and that the level of sophistication is more closely related to students' achievements in a mathematics tests than the belief position.

### 6.1.2 Theoretical Background

Beliefs are commonly recognized as "[p]sychologically held understandings, premises, or propositions about the world that are thought to be true." (Philipp 2007, p. 259). They are thought to be more cognitive and more stable than attitudes or emotions and to filter perceptions, influence learning processes, and direct actions (ibid.; see also Thompson 1992). Research strongly indicates that beliefs are not held solitarily but are organized in clusters around particular ideas or objects, called *belief systems* (Green 1971; Philipp 2007).

Epistemology is a branch of philosophy dealing with the nature of human knowledge, including its limits, justifications, and sources (cf. Arner 1972, ch. I).

In mathematics education, beliefs about the nature of mathematics and of mathematical knowledge are often described as *mathematical worldviews* (Schoenfeld 1985; Törner and Pehkonen 1999). In psychology and education, research on beliefs about the nature of knowledge and knowing—so called epistemological beliefs—and their development are often covered under the term *personal epistemology* (Hofer and Pintrich 1997). Staples et al. (2012) elaborate on the fact that teachers' epistemological beliefs, for example those regarding the justification of knowledge, also shape their behavior in lessons.

In educational psychology, early studies modeled personal epistemology as a one-dimensional sequence of stages (Hofer and Pintrich 1997), whereas nowadays most researchers differentiate between several dimensions to examine epistemological beliefs in detail (Hofer 2000). A widely accepted structure for such a system was proposed by Hofer and Pintrich (1997). It consists of two general dimensions (*nature of knowledge* and *nature or process of knowing*) with two respective sub-dimensions: *certainty* and *simplicity of knowledge* as well as *source* and *justification of knowledge*. These works focus on elaboration and context specificity of epistemological (and ensuing pedagogical) beliefs.

However, there are **theoretical issues**, as the structure (e.g., the dimensionality) and the interpretation of these constructs are considered controversial. Also, it remains unclear to what level epistemological beliefs are general or domain-specific (Hofer 2000) and why these beliefs seem to be less stable than hitherto assumed (cf. Greene et al. 2008; Stahl 2011). To overcome these theoretical issues, Stahl (2011)—drawing on theories of cognitive flexibility (Jacobson and Spiro 1995)—suggested theoretically differentiating between relatively stable *epistemological beliefs* and situation-specific *epistemological judgments*. The latter are defined

> [...] as learners' judgments of knowledge claims in relation to their beliefs about the nature of knowledge and knowing. They are generated in dependency of specific scientific information that is judged within a specific learning context. [...] [A]n epistemological judgment might be a result of the activation of different cognitive elements (like epistemological beliefs, prior knowledge within the discipline, methodological knowledge, and ontological assumptions) that are combined by a learner to make the judgment. (Stahl 2011, p. 38 f.)

Notice that this definition is different from the distinction of central and peripheral beliefs according to Philipp (2007) since it relies on the contextual flexibility rather than on amenability to change. Therefore, it seems fruitful to investigate such judgments and their dependency on the information and learning context.

Alongside these theoretical issues, there are **methodological issues** regarding the instruments with closed question formats as used by a majority of the studies in both mathematics education and psychology (Duell and Schommer-Aikins 2001). For example, in mathematics education, the COACTIV study (Baumert et al. 2009, pp. 63 ff.), building on Grigutsch et al. (1998), used questions with a four-point Likert scale (is not correct/is rather not correct/is rather correct/is correct). Example items are: "Mathematical theorems exist a priori. However, they have to be discovered." or "Very important for mathematics is its logical rigor and precision, that is its 'objective' thinking."

In psychological research, the most common method of measuring epistemological beliefs is through the use of questionnaires that build on Schommer's (1990) questionnaire, which also uses Likert scale items (Hofer 2000). Stahl (2011, p. 41 f.) states that there has been little success in developing a questionnaire with strong reliability and validity. He identifies the main problem in the unstable factor structure of the instruments and sees another problematic aspect in items, which are often indirectly related to epistemological beliefs. Muis (2004) identifies additional difficulties with questionnaires in their effectiveness and in their capability of measuring general as well as domain-specific epistemological beliefs.

To evaluate the categories and instruments used in research on personal epistemology, Greene and Yu (2014) conducted an interview study with experts (university professors) from the academic domains of biology and history as well as novices (middle school students). They (ibid., p. 23) "identified a number of ways that models of EC [epistemic cognition] might be profitably altered to better match novice and expert beliefs about the natures of knowledge and knowing [as well as] ways that the measurement validity of EC self-report instruments might be improved." For example, Greene and Yu identified differences between procedural and declarative knowledge claims (e.g., important dates like signing of the Declaration of Independence that just need to be remembered).

> Therefore, models that do not account for differences in the "simple" and "certain" nature of knowledge beliefs across types of knowledge, and instruments that include items about "knowledge" in general, are unlikely to adequately discriminate between novices and experts, leading to poor measurement validity. (Greene and Yu 2014, p. 23)

To overcome these theoretical and methodological issues, we differentiate between an individual's belief position and his or her argumentation to justify the belief position. For example, a person might hold the position that mathematical knowledge is certain and he or she might have more sophisticated arguments (mathematical rigor, deductive reasoning, an elaborate review system, ...) or less sophisticated arguments ("If any knowledge is certain, it is mathematical knowledge. Didn't you learn this in school?") to back up his or her position.

In psychological research, the degree to which a belief is rated as naïve or sophisticated is often considered to be determined by the assessed belief position. For example, a belief about the certainty of knowledge is considered as naïve or inflexible, when it expresses agreement to knowledge claims about immutability and certainty of knowledge (e.g., Schommer 1998). On the other hand, a belief is considered as sophisticated when it expresses agreement to knowledge claims about changeability and uncertainty of knowledge (ibid.).

We believe that at least for the domain of mathematics, the belief position is not tied to its sophistication in such a strong way. For the area of "certainty of knowledge", we were able to show in an interview study (Rott et al. 2014) that both positions—*mathematical knowledge is certain* vs. *uncertain*—can be held with sophisticated as well as naïve and inflexible arguments. We were then able to develop an extended questionnaire with open-ended items and to show that belief position (*certain* vs. *uncertain*) and argumentation (*inflexible* vs. *sophisticated*) were indeed unrelated in a sample of more than 200 university students (Rott et al. 2015). Additionally, only the argumentation but not the position was correlated to a test of mathematical critical thinking. Students who argued sophisticatedly scored significantly better than students who argued inflexibly in the critical thinking test (ibid.).

The goal of this paper is to validate these findings by extending them to another area of epistemological beliefs, namely the *justification of knowledge*, which is "the central question of philosophical epistemology" (Greene et al. 2008, p. 146).

### 6.1.3 Justification of Knowledge

The dimension *justification*, which is frequently addressed in research on personal epistemology, deals with the way in which individuals evaluate knowledge claims by drawing on evidence and authority (Hofer 2000, p. 380). It also describes how individuals determine which claims are sufficiently justified to be considered knowledge, for example by the use of analytic tools such as sourcing (Greene et al. 2008, p. 142).

As a focus of our study, we chose to specify the epistemological dimension of justification with regard to the process of *discovery in mathematics*, thus introducing a domain specific perspective. To explain our focus and to discuss the potential flexibility of epistemological judgments, we take a closer look into the complementary role of induction and deduction as core processes of discovery

(Boero 1999). (The further differentiation of inductive processes into induction and abduction, as introduced by Peirce et al. 1960, is neglected throughout the following argument.) Although deductive proof is generally regarded as a central and unique characteristic of mathematics, this is only half the truth. Pólya (1954) pointed out that the discovery of a hypothesis mostly relies on *inductive* methods like using and testing examples, drawing figures, stating provisional hypotheses, and so on.

Although mathematicians usually do not publish conjectures, there are unproven statements such as Goldbach's conjecture or the Riemann hypothesis which are regarded as belonging to the body of mathematical knowledge. Pólya (1954) also presents the case of Leonard Euler (e.g., Euler 1761) who published a conjecture being "assured of its truth without giving it a perfect demonstration. Nevertheless, I shall present such evidence for it as might be regarded as almost equivalent to a rigorous demonstration." (Pólya 1954, p. 91). This short sketch on the double-faced quality of the generation of knowledge in mathematics—omitting the social dimension (e.g., Lakatos 1976) for sake of brevity—hints at the possibility to hold each view—"mathematical knowledge is justified *inductively*" versus "mathematical knowledge is justified *deductively*"—with sophisticated arguments.

Our **research intentions** address both the theoretical and methodological issues of research in personal epistemology. We are looking for empirical confirmation of the complexity of epistemological beliefs in the sense described above with respect to the area of justification of knowledge. We are also looking for empirical results that either corroborate or question the validity of closed-ended instruments commonly used in research on epistemological beliefs such as the degree of sophistication with which a belief is being held.

To address these research intentions, two studies have been carried out, a qualitative interview study as well as a quantitative questionnaire study.

## 6.2 Study 1

To investigate epistemological beliefs regarding the justification of (mathematical) knowledge, we conducted an interview study (see Chap. 2). The cases were selected with respect to expected breadth of personal epistemologies. New cases were added until a certain degree of saturation was reached (cf. Mayring 2000). To ensure a certain range of arguments, not only students but also teachers and teacher educators were asked to participate. The varying familiarity of these groups with mathematical reasoning and the varying contexts in which these groups reflect upon mathematical epistemology are expected to lead to a more pronounced variety of arguments. However, we do not intend to draw inferences with respect to the cause of such differences beyond the group of pre-service teachers.

Overall, 17 participants were interviewed by the first author. All participated voluntarily: 10 student teachers in pre-service education and two teachers of mathematics with more than 10 years of teaching experience each, two professional mathematicians as well as two professors of mathematics and one professor of mathematics education (see Table 2.1 in Chap. 2). In addition to being interviewed, these participants filled out a short questionnaire with Likert scale items that were derived from the questionnaire by Grigutsch et al. (1998). Three of those items were related to the justification of knowledge and are shown in Fig. 6.1.

To evoke the possible context dependency, we used complementary and even contradicting situations, framing the subjects' utterances with respect to their personal epistemology. At the start of the semi-structured interviews, we confronted our participants with two quotes representing a deductive and an inductive position respectively (see Table 6.1) and asked them to react to the prompt: "These are two positions of mathematicians regarding mathematical discovery. With which position can you identify? Please give reasons for your answer." This

| Mathematics as a scientific discipline from my point of view | | | | | | |
|---|---|---|---|---|---|---|
| New mathematical theories evolve only when a number of statements is combined with a (flawless) proof. | A.H. | ① | ✖ | ③ | ④ | ⑤ |
| | D.B. | ① | ✖ | ③ | ④ | ⑤ |
| | B.G. | ① | ② | ③ | ✖ | ⑤ |
| | C.P. | ① | ② | ③ | ✖ | ⑤ |
| Developing mathematical theories, one has to accept errors; crucial are good ideas. | A.H. | ① | ② | ③ | ④ | ✖ |
| | D.B. | ① | ② | ③ | ✖ | ⑤ |
| | B.G. | ✖ | ② | ③ | ④ | ⑤ |
| | C.P. | ✖ | ② | ③ | ④ | ⑤ |
| Flawlessness is only demanded for logical deductions, not in its development. | A.H. | ① | ② | ③ | ✖ | ⑤ |
| | D.B. | ✖ | ② | ③ | ④ | ⑤ |
| | B.G. | ✖ | ② | ③ | ④ | ⑤ |
| | C.P. | ✖ | ② | ③ | ④ | ⑤ |

① strongly agree … ⑤ strongly disagree

For comparison, our categorization from above:

| Mathematical discovery is justified | **Inflexible** | **Sophisticated** |
|---|---|---|
| **deductively** | A.H. | D.B. |
| **inductively** | B.G. | C.P. |

**Fig. 6.1** Interviewees' reactions to the questionnaire, items regarding justification

**Table 6.1** Starting positions for "mathematical discovery."

| Mathematical discovery is justified deductively | Mathematical discovery is justified inductively |
|---|---|
| "Mathematics is a deductive science: starting from certain premises, it arrives, by a strict process of deduction, at the various theorems which constitute it. [...] No appeal to common sense, or 'intuition,' or anything except strict deductive logic, ought to be needed in mathematics after the premises have been laid down." (Bertrand Russell) [1919, p. 144] | "Mathematics is an experimental science. The formulation and testing of hypothesis play in mathematics a part not other than in chemistry, physics, astronomy, or botany. [...] It matters little that the mathematician experiments with pencil and paper while the chemist uses testtube and retort, or the biologist stains and the microscope." (Norbert Wiener) [1923, p. 237 ff.] |

proved to generate more insight into our subjects' beliefs than direct questioning (e.g., "What constitutes mathematical discovery?"), as we could ascertain in pilot interviews. Using the statements to initiate the interviews, our interviewees had different positions to refer to and got an impression of the breadth of the argument.

After the initial prompt, we posed additional questions and intervened with information contrary to the subjects' positions to further identify their lines of reasoning: "Can you explain your position on the basis of your mathematical experience?", "Please compare the discovery of mathematical knowledge to that of other sciences, for example to knowledge in physics, linguistics, or educational science." If a subject settled on "mathematical discovery is deductive", we confronted him/her with Goldbach's conjecture, which is widely accepted but still unproven. If the interviewee committed on "mathematical discovery is inductive", we showed a formula by Euler to demonstrate the importance of proofs. (This formula generates prime numbers up to $n = 40$ but fails to do so for $n \geq 41$.)

The interviews have been analyzed by a team of researchers. Within extensive group discussion, the results have been validated consensually following Hill et al. (1997, p. 523): "Team members first examine the data independently and then come together to present and discuss their ideas until they reach a single unified version that all team members endorse as the best representation of the data. Using several researchers provides a variety of opinions and perspectives, helps to circumvent the biases of any one person, and is helpful for capturing the complexity of the data." During our discussions, there were no disagreements regarding the interviewees' belief positions but longer arguments about their belief justifications. We present only a selected sample of interviews.

### 6.2.1 Results of the Interview Study

We present the results of the analysis of the interview data with respect to the "justification of mathematical discovery" in two steps: First, we outline the breadth of the arguments of our interviewees, who can be separated into two groups (preferring either "deductive" or "inductive"). Second, we present the subjects' lines of reasoning to illustrate that subjects, who support the same position and choose the same responses in a typical beliefs questionnaire, can do so for differing reasons and with different degrees of sophistication. The excerpts presented here are drawn from the data for reasons of typicality. We want to show representatives of all four possible outcomes of position ("deductive" vs. "inductive") combined with argumentation ("inflexible" vs. "sophisticated"). We deliberately chose examples from the non-experts to show that even (pre-service) teachers can hold sophisticated beliefs. Additionally, we present the summary of one of the expert's interviews, who refused to commit himself to a belief position. Finally, we summarize the interpretation of this evidence as support for our theoretical assumption of flexibility of epistemological judgments.

We start with interviewees judging that "mathematical discovery is justified *deductively*".

(1) A.H. is a pre-service teacher in her second year at a university in South Germany (University F) (age 22). She states that for her, something can only be recognized as mathematical discovery "if it's really been proven." As long as a discovery is only relying on examples—"being in an experimental stage"—she would rather call it "working hypothesis". Being asked what she thinks about scientific journals that deliberately publish unproven hypotheses, A.H. responds that these hypotheses wouldn't count as discoveries for her. She would rather interpret these publications as a request for mathematicians to look for a proof. She doubts such hypotheses and wouldn't build anything on them. Finally, according to the pre-defined interview structure, the interviewer asks her about her thoughts on Goldbach's conjecture. This conjecture hasn't been proven but it has been confirmed by countless examples; wouldn't such a confirmation be more convincing than a proof that almost nobody can comprehend? A.H. answers that personally she isn't satisfied without a proof. Examples might make it more accessible to most people, but for her, the consideration of a lot of examples would better fit to physicists.

(2) D.B. is a secondary teacher in North Germany with more than 10 years of teaching experience in mathematics and physics (age 42). He declares that creativity is an extremely important part of mathematics and working the

way Wiener describes it is essential for it. But in the end "you have to write it down precisely and rigorously in a deductive way." According to D.B., one missing step would spoil an entire proof. In comparison to mathematicians, he continues, physicists can really work inductively. They have to design experiments and build hypotheses on them. But, rigorously, physicists are unable to say whether their discoveries hold true in another solar system. A mathematical theorem, on the other hand, is true even in solar systems that are not ours. Regardless of the number of times a mathematical experiment is carried out—e.g. in stochastics—a hypothesis has to be proven deductively. Even Euler, who had published unproven results, advocated this position and further worked on proofs. The same was true for mathematicians such as Riemann or Gauss who had worked with calculations, sketches and examples on countless pages of their research diaries. "Finally, mathematical discovery needs a systematic, complete, deductive proof."

We continue with interviewees judging that "mathematical discovery is justified *inductively*":

(3) B.G. is a pre-service teacher who just finished her degree at the University F (age 25). After clarifying the terms "deductive" and "inductive", she states that mathematicians work inductively. She says that she cannot imagine mathematicians solely using a deductive line of reasoning to arrive at their discoveries. For her, writing down a deductive proof is only the final part of an argumentation. But getting to mathematical discovery and even justifying it is an inductive process. She was not able to support her position with further arguments, though.

(4) C.P. is a pre-service teacher in her third year at the University F (age 23). She states that working mathematically is an interaction of inductive and deductive processes. Based on observations, mathematicians would gain their hypotheses, and later on they would try to connect these to already known theorems. Asked about "mathematical discovery", C.P. answers that even unproven hypotheses could count as such. It would be sufficient for such a hypothesis to initiate mathematical discourse. In such a discourse, several people would think about the hypothesis and whether or not it is true. She uses the four-color theorem to illustrate her point of view. This theorem had been a "mathematical discovery" long before it got accepted as proven. Even counter-examples wouldn't disqualify a hypothesis as being a discovery. Such examples could lead to a deeper understanding of the mathematics behind the

hypothesis and could initiate an adapted version of it that is "a bit closer to the truth".

We present an excerpt from a fifth interview, as a representative of the experts.

(5) S.W. is a professor for mathematics at a university in North Germany. He notices that the two positions by Russell and Wiener appear to be contradicting at first, but are not necessarily so. Wiener describes the way that mathematical knowledge is created, whereas Russell describes the way in which it is written down. A good mathematician, S.W. continues, should be able to integrate both positions into his or her work. But it is his experience that a lot of students—even some of those who want to write a Ph.D. thesis in mathematics—refer only to the position of Russell. Those are the students who are missing creativity and intuition. A comparison would be someone who plays the piano perfectly but is not able to compose music like Bach.

### 6.2.2 Interpretation with Respect to the Category of Epistemological Judgments

These summarizing excerpts from our interview study support the theoretical differentiation between epistemological beliefs and judgments: Both A.H. and D.B. answered the question whether mathematical discovery is justified deductively or inductively by choosing "deduction". This belief position was also reflected by their reaction to the questionnaire part as both answered in favor of deduction accordingly (see Fig. 6.1). Nevertheless, their lines of reasoning differ substantially. A.H. states that she is only satisfied having a proof, but she is not able to give any reasons for that other than personal preferences. In contrast, D.B. argues more sophisticatedly by pointing out that an argumentation is spoilt by missing even one step. He highlights the general validity of deductively proven statements ("such a theorem [...] would hold true in a solar system other than ours"). He also compares discovery in mathematics and in physics and connects his position to the work of famous mathematicians.

An analogous situation can be found in comparing B.G. and C.P., who both stated that "mathematical discovery is justified inductively". They also showed matching responses in favor of "induction" in the questionnaire (see Fig. 6.1). However, in the interviews, they expressed considerably differing argumentative backings. B.G., on the one hand, had unreflected beliefs and was only able to say that she could not imagine mathematicians getting to their results by only reasoning deductively. C.P., on the other hand, showed very sophisticated beliefs and was able to give reasons (mathematical discourse) and examples (four color theorem) for her position. She was even able to defend her position against the possibility of wrong hypotheses.

Therefore, we conclude that the empirical data represented by these examples actually support the theoretical claim of Stahl (2011): Questionnaires with rating scales (like the short one we used) can differentiate only between belief positions, but not between (more or less sophisticated) belief argumentation:

> In a questionnaire with rating scales, [these] persons would give the same answer. However, the conclusion that their responses are an expression for comparable epistemological beliefs would be wrong. Their epistemological judgments are built on different cognitive elements to evaluate the knowledge claim. (Stahl 2011, p. 49)

The mathematician S.W. shows very sophisticated beliefs regarding mathematical discovery and is able to integrate both positions into his statements. We will pick up this observation in the discussion section of this article

An overview of the interpretations of the 17 interviews is given in Fig. 6.2. Most participants chose one of the two belief positions. However, some experts, like S.W., argued that both positions are applicable, depending on the situation at hand.

| Mathematical discovery is justified | **inflexible** | **sophisticated** |
|---|---|---|
| **deductively** | K.N. (pre-service teacher)<br>A.H. (pre-service teacher)<br>T.H. (pre-service teacher)<br>C.K. (pre-service teacher)<br>D.S. (pre-service teacher)<br>H.K. (pre-service teacher) | D.B. (secondary teacher)<br>A.R. (mathematician)<br>A.S. (mathematician) |
| **(refused to choose between deductively and inductively)** | | R.E. (secondary teacher)<br>K.E. (economics professor)<br>S.W. (math professor)<br>T.B. (math ed professor) |
| **inductively** | B.G. (pre-service teacher)<br>T.W. (pre-service teacher) | C.P. (pre-service teacher)<br>P.S. (pre-service teacher) |

**Fig. 6.2** Interviewees' mapping to the dimensions belief position and argumentation

## 6.3 Study 2

In study 1, we were able to show that differences in belief position and belief sophistication can be identified by the use of interviews. Based on the results of study 1, we developed an extended questionnaire with open items to be able to capture both, positions and argumentations, within the answers to epistemological questions. By avoiding Likert scale items, we tried to take into account concerns and suggestions expressed by Muis (2004), Stahl (2011), as well as Greene and Yu (2014). To stimulate the participants to write extensive answer texts, we used the same quotes (see Table 6.1) and questions as in the interview guideline.

In contrast to study 1, which included interviewees from different groups (students, teachers, mathematicians), we concentrated on pre-service teachers in study 2.

At the beginning of the winter semester of 2014/15, we presented this questionnaire within lectures (or related tutorials) for pre-service mathematics teachers from a university in South Germany (University of Education, Freiburg) (teachers for primary and lower secondary schools) and a university in West Germany (University of Duisburg-Essen) (teachers for lower and upper secondary schools). Those lectures were designed to convey mathematical content (like basic arithmetic) without addressing epistemological beliefs explicitly. The students were asked to fill in the questionnaires voluntarily in the lecture time. Participating in the study or refusing to do so did not affect the students' outcome of those lectures. 467 students participated in this study, of which 439 wrote answers that were long enough to enable coding regarding belief position and argumentation. This led to the following numbers of pre-service teachers: $n = 194$ for primary schools (all from the university in Freiburg), $n = 145$ for lower secondary schools ($n = 76$ from the university in Freiburg and $n = 69$ from the university in Essen), and $n = 100$ for upper secondary schools (all from the university in Essen). Examples for each of the four theoretically possible outcomes are shown in Table 6.2.

The students' answers were coded independently by two trained raters. The reliability score shows a very high agreement (Cohen's $\kappa = 0.88$). A closer look into the interrater matrix (Table 6.3) reveals that coding for belief position was done without any disagreement, whereas belief argumentation was harder to rate. After calculating the interrater reliability, the differing scores have been re-coded by the two raters consensually (see Table 6.4 for the final codes).

In addition to the belief questions, we ran a short mathematical test with items similar to the well-known bat-and-ball task by Kahneman and Frederick (2002): "A bat and a ball cost $ 1.10 in total. The bat costs $ 1 more than the ball. How much does the ball cost?" All of those items were designed (a) to be solvable

**Table 6.2** Examples for pre-service teachers' answers in the questionnaire on "mathematical discovery."

| | Inflexible | Sophisticated |
|---|---|---|
| Deductively | Of course, mathematicians conduct experiments, but theorems are not derived from experiments, those can only be used for illustration. [...] (MMI-18) | Experimenting plays a major role in discovering new phenomena. But these have to be justified deductively by going back to a strong mathematical foundation before they can be called mathematical knowledge. [...] (EYNZ-15) |
| Inductively | Mathematical discoveries are figured out and determined by trial and error and experimenting. [...] (WMAI-04) | There are underlying premises, but mathematical discoveries are justified inductively by trial and error and experimenting before mathematicians even try to prove them. [...] (WSEL-16) |

**Table 6.3** Calculation of the interrater-reliability coding "mathematical discovery"

| | Rater 1 | | | | | |
|---|---|---|---|---|---|---|
| Rater 2 | | Deductive & Inflexible | Deductive & Sophisticated | Inductive & Inflexible | Inductive & Sophisticated | Sum |
| | Deductive & Inflexible | 130 | 9 | 0 | 0 | 139 |
| | Deductive & Sophisticated | 4 | 10 | 0 | 0 | 14 |
| | Inductive & Inflexible | 0 | 0 | 260 | 9 | 269 |
| | Inductive & Sophisticated | 0 | 0 | 7 | 10 | 17 |
| | Sum | 134 | 19 | 267 | 19 | 439 |

$P_{\text{obs}} = 0.934$, $P_{\text{exp}} = 0.472$, Cohen's $\kappa = 0.875$

with simple mathematical procedures, but allow (b) to suggest an intuitive but incorrect answer like $ 0.10 in the bat-and-ball task. With these items we assessed the reflection on the solutions and critical thinking in problem solving. After validating the items in preliminary studies (Rott and Leuders 2016a, b), the final test consisted of 11 items that were rated dichotomously.

**Table 6.4** Comparison of both dimensions of the extended questionnaire for all students; the numbers in parentheses indicate expected frequency under the assumption of statistical independence

| | Inflexible | Sophisticated | Sum |
|---|---|---|---|
| Deductively | 140 (141.8) | 13 (11.2) | 153 |
| Inductively | 267 (265.2) | 19 (20.8) | 286 |
| Sum | 407 | 32 | 439 |
| $\chi^2 = 0.27$; df = 1; $p = 0.603$ | | | |

We used a Rasch model to transform the test scores into values on a one-dimensional competence scale (software RUMM 2030 by Andrich et al. 2009), called *mathematical critical thinking.* We had to eliminate two items because of underdiscrimination (fit residual > 2.5) in connection with floor and ceiling effects, respectively. After that elimination, for each item the model showed good fit residuals (all values between –2.5 and 2.5) and no significant differences between the observed overall performance of each trait group and its expected performance (overall-$\chi^2$ = 36.2; df = 27; p = 0.11).

## 6.3.1 Results of the Extended Questionnaire Study

The results of the consensual re-coding of the students' belief positions and argumentations are shown in Table 6.4. A chi-square test was used to investigate whether the two test dimensions—position (*deductive* vs. *inductive*) and argumentation (*inflexible* vs. *sophisticated*)—are related. As expected, there is no relation between those two dimensions ($\chi^2 = 0.27$, $p = 0.603$). This finding strongly suggests that the sophistication of a belief position cannot reliably be measured by Likert scale items that primarily capture belief positions.

The distribution of the pre-service teachers' epistemological beliefs and their degree of sophistication is presented in Table 6.5—sorted by their number of semesters (median split) and by their aspired teaching profession (primary, lower secondary, or upper secondary school).

Surprisingly, there was no increase in the percentage of sophisticated positions with a growing number of semesters. In both groups—students of three or less and students of four or more semesters—the same proportion of students (approximately 7.3%) argued sophisticatedly. We had expected an improvement in the argumentation with growing experience in content knowledge and pedagogical

**Table 6.5** Distribution of the pre-service teachers' beliefs and their degree of sophistication

| | Deductive & Inflexible | Deductive & Sophisticated | Inductive & Inflexible | Inductive & Sophisticated | Sum |
|---|---|---|---|---|---|
| Semester ≤ 3 | 70 (30.0%) | 7 (3.0%) | 146 (62.7%) | 10 (4.3%) | 233 (100%) |
| Semester ≥ 4 | 70 (34.0%) | 6 (2.9%) | 121 (58.7%) | 9 (4.4%) | 206 (100%) |
| Primary | 54 (27.8%) | 3 (1.5%) | 127 (65.5%) | 10 (5.2%) | 194 (100%) |
| Lower secondary | 47 (32.4%) | 3 (2.1%) | 91 (62.8%) | 4 (2.8%) | 145 (100%) |
| Upper secondary | 40 (40.0%) | 6 (6.0%) | 49 (49.0%) | 5 (5.0%) | 100 (100%) |
| All students combined | 140 (31.9%) | 13 (3.0%) | 267 (60.8%) | 19 (4.3%) | 439 (100%) |

content knowledge. This result suggests that questions regarding the justification of mathematical knowledge are not covered in the students' courses.

The students' mean scores on the mathematical critical thinking test are presented in Table 6.6. The Rasch model of the students' critical thinking ability provides metrical latent variables ranging from –2.83 to 2.81, with low values indicating a low ability. The data are sorted by the number of semesters (rows

**Table 6.6** Means (and standard deviations) of "mathematical critical thinking."

| | Deductive | Inductive | Inflexible | Sophisticated | Total |
|---|---|---|---|---|---|
| Semester ≤ 3 | −0.22 (0.94) | −0.04 (0.84) | −0.15 (0.84) | 0.58 (1.08) | −0.10 (0.88) |
| | n = 77 | n = 156 | n = 216 | n = 17 | n = 233 |
| Semester ≥ 4 | −0.03 (1.07) | −0.17 (0.99) | −0.19 (0.99) | 0.79 (1.03) | −0.12 (1.02) |
| | n = 76 | n = 130 | n = 191 | n = 15 | n = 206 |
| Primary | −0.26 (0.96) | −0.15 (0.89) | −0.26 (0.86) | 0.80 (1.06) | −0.18 (0.91) |
| | n = 57 | n = 137 | n = 180 | n = 14 | n = 194 |
| Lower secondary | −0.35 (0.98) | −0.17 (0.90) | −0.27 (0.91) | 0.60 (1.14) | −0.23 (0.93) |
| | n = 50 | n = 95 | n = 138 | n = 7 | n = 145 |
| Upper secondary | 0.28 (1.01) | 0.16 (0.95) | −0.16 (0.96) | 0.59 (1.03) | 0.21 (0.97) |
| | n = 46 | n = 54 | n = 89 | n = 11 | n = 100 |
| All students combined | −0.13 (1.01) | −0.10 (0.91) | −0.17 (0.91) | 0.68 (1.04) | −0.11 (0.95) |
| | n = 153 | n = 286 | n = 407 | n = 32 | n = 439 |

two and three) and by the students' aspired teaching profession (rows four, five, and six) as well as by their belief positions (columns three and four) and their argumentation (columns five and six).

A t-test was used to compare the two groups that aspire to teach at lower secondary schools (University Freiburg: –0.228 (0.873), $n = 76$; University Essen: –0.231 (0.998), $n = 69$). The test showed no significant differences between these two groups ($t = 0.02$, df $= 143$, $p_{\text{two-tailed}} = 0.984$), so that we combined these two groups in our analyses.

Does the level of mathematical critical thinking differ between students of a different belief position (*inductive* vs. *deductive*)? A two-way ANOVA has been used to investigate possible differences between students' critical thinking scores sorted by their position: There is no significant difference between students regarding knowledge justified "inductively" compared to students regarding it justified "deductively" ($F = 0.09$; $p = 0.764$). There are, however, significant differences between students of the different study programs ($F = 7.69$; $p = 0.0005$); a Tukey Post Hoc test reveals that these differences are significant between primary and upper secondary as well as between lower and upper secondary ($p < 0.01$), but not between primary and lower secondary ($p > 0.05$) pre-service teachers. There is no interaction effect ($F = 1.01$; $p = 0.365$). These results are in accordance with our expectations: On the one hand, we did not expect the belief position to interact with critical thinking scores. On the other hand, teachers for upper secondary schools generally score better in mathematics-related tests than those for primary and lower secondary schools.

Does the level of mathematical critical thinking correspond to the degree of sophistication of the epistemological judgments? A second two-way ANOVA has been used to analyze the critical thinking scores sorted by the quality of the argumentation. This test confirms the significant differences between the three study programs ($F = 8.10$; $p = 0.0004$). Additionally, students arguing sophisticatedly show higher ability scores than students arguing in an inflexible way ($F = 25.68$; $p < 0.0001$). There is no interaction effect ($F = 0$; $p = 1$). Thus, as expected, there is a marked connection between sophistication of beliefs and the ability of thinking critically when solving tasks.

A third ANOVA was used to test whether students from higher semesters (number of semesters $\geq 4$) score better than students from lower semesters (number of semesters $\leq 3$) in the critical thinking test. This test also confirms the differences between the three study programs ($F = 7.95$; $p = 0.0004$). Surprisingly, there is no effect for the number of semesters ($F = 0.05$; $p = 0.823$). Possibly, since the tasks of the critical thinking test do not require any knowledge past school level, they do not distinguish between students of higher or lower number of semesters. This result for the whole population, which is not in accordance with our expectations, can also be caused by the significant interaction effect of study program and number of semesters ($F = 8.40$; $p = 0.0003$). In the group of pre-service teachers for upper secondary schools, older students score better. However, there are no differences in the "lower secondary" group; and in the "primary" group, older students score even worse than students with three or less semesters.

## 6.4 Discussion

The starting point of the investigations presented in this report was the state of research in capturing epistemological belief systems. Apart from the heterogeneity of theories (belief as a "messy construct", Pajares 1992) and the lack of integration of the different strands of research, there is a general dissatisfaction with the inconclusive empirical evidence for certain belief structures. Therefore, it seemed appropriate to investigate beliefs from a slightly different angle by distinguishing between a belief position and the quality of the argumentation for a belief position. In this context, it was reasonable to follow a theoretical argument that pleads for taking into account the flexibility and the context dependence of personal epistemologies.

In previous studies (Rott et al. 2014, 2015) on epistemological beliefs, this approach turned out to be fruitful. But since these results focused only on the area of certainty of knowledge, we intended to look for additional empirical backing in a different area. In this paper, we presented two studies focusing on the subdimension justification of knowledge.

### 6.4.1 Discussion of Study 1

Within the framework of the flexibility and the context dependence of personal epistemologies, it seemed plausible to capture epistemological judgments in a semi-structured interview stimulating varied contexts (instead of pre-structured questionnaires). The qualitative data collected actually produced evidence for using such a differentiated approach to capture the assumed flexibility. Within our group of interviewees, all four theoretically possible combinations of belief position (mathematical knowledge is justified *deductively* or *inductively*) and argumentation (the position can be backed up by either *inflexible* or *sophisticated* arguments) were identified. So far, this had been only supported by hypothetical examples (Stahl 2011, p. 52 ff.).

The beliefs that we focus on in this study are strongly related to acting and thinking mathematically. Therefore, we chose to integrate experts on the epistemology of mathematics (experienced teachers and teacher educators) as participants into the interview study. This provided us with the opportunity to fully explore the potential width of arguments supporting different belief positions that could not be expected from students. Even though we found sophisticated arguments within the group of pre-service teachers, the integrating position that argued for both belief positions was only shown by the experts (see interview #5 by S.W.)

The results of this study clearly indicate that holding a certain belief position cannot be identified with being able to justify that position more or less sophisticatedly. Instead, belief position and argumentation should be collected independently. This was taken into account for the development of the extended questionnaire we used in study 2.

### 6.4.2 Discussion of Study 2

In study 2, we used an extended questionnaire in a sample of more than 450 pre-service mathematics teachers. Instead of Likert-scale items that have recently been criticized for methodological issues (Muis 2004; Stahl 2011; Greene and Yu 2014), we used open-ended items based on the interview guide from study 1. Similar to study 1, the pre-service teachers all opted for one of the two belief positions; no participant argued for an integration of both positions, even though the open text production clearly provided the opportunity to do so.

We were able to show that both position and argumentation of epistemological beliefs regarding the justification of knowledge were independent and that only the argumentation is correlated to a test measuring critical thinking in mathematics. These results illustrate the importance of measuring epistemological beliefs in a finer grained way than only collecting positions as is often done in commonly used questionnaires (e.g., Schommer 1990).

### 6.4.3 General Discussion

In conclusion, it can be stated that the research into personal epistemologies and their argumentative structure is still an open field for theoretically and methodologically new approaches and insights. In this area, our approach of describing epistemological judgments and their context dependency reveals some promising findings, especially regarding issues of flexibility versus stability of cognitive processes.

Our results have clear implications for research: Many studies (like COACTIV, Krauss et al. 2008) use teachers' subjective beliefs (which include epistemological beliefs) as a covariate. In these studies, those beliefs are generally collected by Likert scale items that identify the participants' position regarding those questions (e.g., knowledge is justified deductively or inductively) (cf. Hofer 2000). Our research shows that rather than the belief position, the belief argumentation (inflexible vs. sophisticated) should be collected and analyzed.

This differentiated view may also prove relevant for further investigations of the impact of belief structures on teachers' classroom decisions. We propose that it could be fruitful to relate mathematics teachers' behavior in theoretical analyses not only to their beliefs held with respect to the nature of knowledge and knowing but also with respect to the sophistication and complexity of their judgements and the source of the warrants given (as e.g. in Nardi et al. 2012).

### 6.4.4 Limitations and Further Studies

There are several limitations that should be improved in future studies. For example, the validity of judging the participants' sophistication is dependent on their willingness to fill out our questionnaire and to fully understand the knowledge claim they are about to judge. The latter limitation concerns a small part of our participants who answered whether the mathematicians work inductively or deductively and not whether mathematical knowledge is justified this way or the other. Those students' answers have been coded as "inflexible". For example:

> Inductively—I think that each mathematician has only come to his conclusions by experimenting and trial and error. […] (SGYR-13)

In future studies, we are going to try different quotes (one from Pólya instead of from Wiener) to start the reasoning in the questionnaire. This way, we aim to reduce the number of students who argue towards the working practices of mathematicians instead of the justification of knowledge as SGYR-13 did. To further ensure the validity of our method, we are going to run a separate test on writing motivation to investigate the first limitation. Additionally, the use of Toulmin's (1958) scheme to analyze arguments might be a fruitful addition to our method of categorizing students' responses to the questionnaire. Within our approach, the participants' arguments have been analyzed only in a quite basic manner to ascertain a *degree of sophistication*. A more differentiated analysis may also yield different argumentation *structures* and help to take into account the following problem: The interpretation of the professors' interviews in study 1 clearly showed that these persons are experts in mathematics as they used very sophisticated arguments. However, their belief positions were not always clearly identifiable as they were unwilling to choose a definite position. Often, they were able to reason for both positions—mathematical knowledge is justified inductively and deductively—depending on the situation. This reaction can be considered as a further level of sophistication and is in line with the argumentation of Gowers (2013), who states that the question whether mathematics is discovered or invented depends on the mathematical object at hand.

This perception indicates that our analysis of sophistication within interview and questionnaire data is restricted to individuals who have not yet reached a status of an "epistemological expert" by extensive experience and profound reflection. For students of mathematics and teacher students, however, our instrument appears to be able to capture variance, structure, and development of central areas of mathematical beliefs. Therefore, the questionnaire will be used for evaluating the impact of teacher education modules in the near future.

## References

Andrich, D., Sheridan, B. E., & Luo, G. (2009). *RUMM2030: Rasch unidimensional models for measurement*. Perth, Australia: RUMM Laboratory.

Arner, D. G. (1972). *Perception, reason, and knowledge—An introduction to epistemology*. London: Scott, Foresman and Company.

Baumert, J., Blum, W., Brunner, M., Dubberke, T., Jordan, A., Klusmann, U., et al. (2009). *Professionswissen von Lehrkräften, kognitiv aktivierender Mathematikunterricht und die Entwicklung von mathematischer Kompetenz (COACTIV): Dokumentation der Erhebungsinstrumente*. Berlin: Max Planck Institute for Human Development.

Boero, P. (1999). Argumentation and mathematical proof: A complex, productive, unavoidable relationship in mathematics and mathematical education. *International Newsletter on the Teaching and Learning of Mathematical Proof, 4*. Retrieved from http://www.lettredelapreuve.org/OldPreuve/Newsletter/990708Theme/990708ThemeUK.html.

Bromme, R., Kienhues, D., & Stahl, E. (2008). Knowledge and epistemological beliefs: An intimate but complicate relationship. In M. S. Khine (Ed.), *Knowing, knowledge and beliefs. Epistemological studies across diverse cultures* (pp. 423–441). New York: Springer.

Buehl, M. M., & Alexander, P. A. (2006). Examining the dual nature of epistemological beliefs. *International Journal of Educational Research, 45*, 28–42.

Buelens, H., Clement, M., & Clarebout, G. (2002). University assistants' conceptions of knowledge, learning and instruction. *Research in Education, 67,* 44–57.

De Corte, E., Op 'T Eynde, P., & Verschaffel, L. (2002). Knowing what to believe: The relevance of mathematical beliefs for mathematics education. In B. K. Hofer & P. R. Pintrich (Eds.), *Personal epistemology; The psychology of beliefs about knowledge and knowing* (pp. 297–320). Mahwah: Lawrence Erlbaum Associates.

Duell, O. K., & Schommer-Aikins, M. (2001). Measures of people's beliefs about knowledge and learning. *Educational Psychology Review, 13,* 419–449.

Elby, A., & Hammer, D. (2010). Epistemological resources and framing: A cognitive framework for helping teachers interpret and respond to their students' epistemologies. In L. D. Bendixen & F. C. Feucht (Eds.), *Personal epistemology in the classroom: Theory, research, and implications for practice* (pp. 409–434). Cambridge: Cambridge University Press.

Euler, L. (1761). Specimen de usu observationum in mathesi pura. (Example of the use of observation in pure mathematics). *Novi Commentarii Academiae Scientiarum Petropolitanae: Anonymos, 6*(6), 185–230.

Fennema, E., & Franke, M. (1992). Teachers' knowledge and its impact. In D. Grouws (Ed.), *Handbook for research on mathematics teaching and learning* (pp. 147–164). New York: Macmillan.

Gowers, T. (2013). Is mathematics discovered or invented? In M. Pitici (Ed.), *The best writing on mathematics* (pp. 8–20). Princeton: Princeton University Press.

Green, T. F. (1971). *The activities of teaching.* New York: McGraw-Hill.

Greene, J. A., Azevedo, R., & Torney-Purta, J. (2008). Modeling epistemic and ontological cognition: Philosophical perspectives and methodological directions. *Educational Psychologist, 43,* 142–160.

Greene, J. A., & Yu, S. B. (2014). Modeling and measuring epistemic cognition: A qualitative re-investigation. *Contemporary Educational Psychology, 39,* 12–28.

Grigutsch, S., Raatz, U., & Törner, G. (1998). Einstellungen gegenüber Mathematik bei Mathematiklehrern. *Journal Für Mathematik-Didaktik, 19*(1), 3–45.

Hill, C. E., Thompson, B., & Williams, E. N. (1997). A guide to conducting consensual qualitative research. *The Counseling Psychologist, 25*(4), 517–572.

Hofer, B. K. (2000). Dimensionality and disciplinary differences in personal epistemology. *Contemporary Educational Psychology, 25*(4), 378–405.

Hofer, B. K., & Pintrich, P. R. (1997). The development of epistemological theories: Beliefs about knowledge and knowing and their relation to learning. *Review of Educational Research 1997, 67*(1), 88–140.

Jacobson, M. J., & Spiro, R. J. (1995). Hypertext learning environments, cognitive flexibility, and the transfer of complex knowledge: An empirical investigation. *Journal of Educational Computing Research, 12,* 301–333.

Kahneman, D., & Frederick, S. (2002). Representativeness revisited: Attribute substitution in intuitive judgment. In T. Gilovich, D. Griffin, & D. Kahneman (Eds.), *Heuristics and biases: The psychology of intuitive judgment* (pp. 49–81). New York: Cambridge University Press.

Krauss, S., Neubrand, M., Blum, W., Baumert, J., Kunter, M., & Jordan, A. (2008). Die Untersuchung des professionellen Wissens deutscher Mathematiklehrerinnen und -lehrer im Rahmen der COACTIV-Studie. *Journal für Mathematik-Didaktik, 29*(3/4), 223–258.

Lakatos, I. (1976). *Proofs and refutations*. Cambridge: Cambridge University Press.

Mayring, P. (2000). Qualitative content analysis. *Forum: Qualitative Social Research 1*(2), Art. 20. Retrieved from http://www.qualitative-research.net/index.php/fqs/article/view/1089.

Muis, K. R. (2004). Personal epistemology and mathematics: A critical review and synthesis of research. *Review of Educational Research, 74*(3), 317–377.

Nardi, E., Biza, I., & Zachariades, T. (2012). 'Warrant' revisited: Integrating mathematics teachers' pedagogical and epistemological considerations into Toulmin's model for argumentation. *Educational Studies in Mathematics, 79,* 157–173.

Pajares, F. M. (1992). Teachers' beliefs and educational research: Cleaning up a messy construct. *Review of Educational Research, 62*(3), 307–332.

Peirce, C. S., Hartshorne, C., & Weiss, P. (1960). *Pragmatism and pragmaticism and scientific metaphysics* (2nd ed.). Cambridge: Belknap Press of Harvard University Press.

Philipp, R. A. (2007). Mathematics teachers' beliefs and affect. In F. K. Lester (Ed.), *Second handbook of research on mathematics teaching and learning* (pp. 257–315). Charlotte: Information Age.

Pintrich, P. R. (2002). Future challenges and directions for theory and research on personal epistemology. In B. K. Hofer & P. R. Pintrich (Eds.), *Personal epistemology: The psychology of beliefs about knowledge and knowing* (pp. 389–414). Mahwah: Lawrence Erlbaum.

Pólya, G. (1954). *Mathematics and plausible reasoning* (Vol. 1). Princeton: Princeton University Press.

Rott, B., & Leuders, T. (2016a). Inductive and deductive justification of knowledge: Flexible judgments underneath stable beliefs in teacher education. *Mathematical Thinking and Learning, 18*(4), 217–286.

Rott, B., & Leuders, T. (2016). Mathematical critical thinking: The construction and validation of a test. In C. Csikos, A. Rausch, & J. Szitányi (Eds.), *Proceedings of the 40th conference of the international group for the psychology of mathematics education* (Vol. 4, pp. 139–146). Szeged: PME.

Rott, B., Leuders, T., & Stahl, E. (2014). "Is mathematical knowledge certain?—Are you sure?" An interview study to investigate epistemic beliefs. *Mathematica Didactica, 37,* 118–132.

Rott, B., Leuders, T., & Stahl, E. (2015). Assessment of mathematical competencies and epistemic cognition of pre-service teachers. *Zeitschrift Für Psychologie, 223*(1), 39–46.

Russell, B. (1919). *Introduction to mathematical philosophy*. London: Allen & Unwin.

Schoenfeld, A. H. (1983). Beyond the purely cognitive: Belief systems, social cognitions, and metacognitions as driving forces in intellectual performance. *Cognitive Science, 7,* 329–363.

Schoenfeld, A. H. (1985). *Mathematical problem solving*. San Diego: Academic Press.

Schommer, M. (1990). Effects of beliefs about the nature of knowledge on comprehension. *Journal of Educational Psychology, 82*(3), 498–504.

Schommer, M. (1993). Epistemological development and academic performance among secondary students. *Journal of Educational Psychology, 85*(3), 406–411.

Schommer, M. (1998). The role of adults' beliefs about knowledge in school, work, and everyday life. In M. C. Smith & T. Pourchot (Eds.), *Adult learning and development—Perspectives from educational psychology*. London: Lawrence Erlbaum.

Stahl, E. (2011). The generative nature of epistemological judgments (Chapter 3). In J. Elen, E. Stahl, R. Bromme, & G. Clarebout (Eds.), *Links between beliefs and cognitive flexibility—Lessons learned* (pp. 37–60). Dordrecht, The Netherlands: Springer.

Stahl, E., & Bromme, R. (2007). The CAEB: An instrument for measuring connotative aspects of epistemological beliefs. *Learning and Instruction, 17*(6), 773–785.

Staples, M. E., Bartlo, J., & Thanheiser, E. (2012). Justification as a teaching and learning practice: Its (potential) multifaceted role in middle grades mathematics classrooms. *The Journal of Mathematical Behavior, 31,* 447–462.

Staub, F., & Stern, E. (2002). The nature of teachers' pedagogical content beliefs matters for students' achievement gains: Quasi-experimental evidence from elementary mathematics. *Journal of Educational Psychology, 93,* 144–155.

Thompson, A. G. (1992). Teachers' beliefs and conceptions: A synthesis of the research. In D. A. Grouws (Ed.), *Handbook of research on mathematic learning and teaching* (pp. 127–146). New York: Macmillan.

Törner, G., & Pehkonen, E. (1999). *Teachers' beliefs on mathematics teaching—Comparing different self-estimation methods—A case study*. Retrieved from www.uni-duisburg.de/FB11/PUBL/SOURCE/AGET_98Final.ps.

Toulmin, S. (1958). *The uses of argument*. Cambridge: Cambridge University Press.

Trautwein, U., & Lüdtke, O. (2007). Epistemological beliefs, school achievement, and college major: A largescale, longitudinal study on the impact of certainty beliefs. *Contemporary Educational Psychology, 32,* 348–366.

Tsai, C.-C. (1998). An analysis of Taiwanese eighth graders' science achievement, scientific epistemological beliefs and cognitive structure outcomes after learning basic atomic theory. *International Journal of Science Education, 20*(4), 413–425.

Wiener, N. (1923). *Collected works: With commentaries*. Cambridge: The MIT Press.

# Summary and Outlook 7

In this final chapter, the work and the results from sub-project #3 of the research project LeScEd are summarized; some general limitations of the study that have not previously been mentioned are discussed as well. Additionally, an outlook is given: Which other research questions can be addressed with the data? How can data generation and analyses be improved? And how can future research projects build on the work done so far?

Section 7.4 has previously published in German language in the proceedings of the 2017 GDM conference (Rott et al. 2017).[1]

## 7.1 Summary

Research on epistemological beliefs, i.e. beliefs on the nature of knowledge, its limits, sources, and justification, has gained greatly in importance in recent years (Hofer and Pintrich 1997). On the one hand, the development of such beliefs at a high level is regarded as an educational goal; on the other hand, there are indications that such beliefs influence actions. Beswick (2012) summarizes several studies in which the influence of epistemological beliefs on the actions of teachers in teaching is demonstrated.

In this book, studies dealing with epistemological beliefs in the domain of mathematics are presented. Particularly, the results of preliminary studies as well

[1] The article was published in the WTM-Verlag (Wissenschaftliche Texte und Medien) which allows its authors to reuse and republish their work without any limitations. It has been translated and slightly adapted by BR. Calculations for this article have been carried out by Jana Groß Ophoff.



B. Rott, *Epistemological Beliefs and Critical Thinking in Mathematics*, Freiburger Empirische Forschung in der Mathematikdidaktik, https://doi.org/10.1007/978-3-658-33539-7_7

as two parts of the main study from sub-project #3 of the research project LeScEd (Learning the Science of Education) are summarized.

Being unsure about the validity of traditional instruments to measure beliefs for the domain of mathematics, and because of the theory of epistemological judgments by Stahl (2011), we conducted a preliminary interview study (Chapter 2). In this qualitative study, interviews regarding epistemological questions (especially concerning the certainty of mathematical knowledge) were held with pre- and in-service mathematics teachers as well as mathematicians and mathematics educators. This variety of participants was chosen to obtain as large an overview as possible of arguments given by novices and experts. These interviews clearly showed that – at least for the domain of mathematics – sophistication of beliefs is not correlated to the belief position a person represents. Contrary to assumptions from psychological research, persons who argue for the certainty of mathematical knowledge do not necessarily hold naïve or unreflected beliefs. Instead, for both of two antithetical positions (mathematical knowledge is certain vs. uncertain), there were interviewees holding beliefs graded from naïve to sophisticated.

In a set of quantitative and qualitative preliminary studies (Chapter 3), a test for mathematical critical thinking (CT) was developed. CT was chosen, because, according to recent competence models, it is a component of students' competencies. The CT test consists of several tasks that are supposedly easy to solve but whose results need to be checked. An exemplary item is the famous bat-and-ball task by Kahneman (e.g., 2011). In several sub-studies, the CT tasks are analyzed and compared to non-CT tasks, showing their quality of revealing a disposition to reflect upon results. Further analyses show that the final version of the CT test can be regarded as a one-dimensional test. Additionally, correlations of CT-test results with mathematical problem solving, mathematical reasoning, and metacognition are explored.

Following the interview study regarding beliefs, we wanted to further explore the independence of belief position and belief argumentation as well as different (inflexible as well as sophisticated) arguments for different belief positions; especially, we wanted to know whether this could be found solely within a group of students without using experts. Additionally, we wanted to investigate the distribution of students arguing sophisticatedly. Therefore, the experience from the interview study was used to develop an open-ended questionnaire (Chapter 4). A quantitative study ($n = 147$) was conducted, in which students' responses were coded for belief position (mathematical knowledge is certain vs. uncertain) and belief argumentation (inflexible vs. sophisticated). The result from the interview study was confirmed: belief position and belief argumentation are statistically independent of each other. Additionally, the CT-test was used in this sample and revealed the following result. Belief position is not correlated to CT-test results, whereas sophisticated belief argumentations are significantly correlated to high CT-test scores. This result further indicates the importance of taking into account belief argumentations instead of only belief positions (as it is done in established instruments to measure epistemological beliefs).

To confirm the results of the questionnaire study, a larger quantitative belief study ($n = 463$), also regarding the topic "certainty of mathematical knowledge", was conducted (Chapter 5). The main results of the preliminary study could all be replicated. Belief positions and argumentations could be coded with great inter-rater agreement and both are statistically independent. Again, CT-test scores are not related to belief position, but highly significantly correlated to belief argumentation. This study also extends the results to more strands of pre-service teacher education (upper secondary teachers in addition to the previously involved groups of primary and lower secondary teachers).

A second part of the larger quantitative belief study extended the research subject to the dimension of "justification of mathematical knowledge" (Chapter 6). In this study ($n = 439$), the main results of the previous belief studies could all be confirmed. Belief position (mathematical discovery is justified deductively vs. inductively) and argumentation (inflexible vs. sophisticated) are independent and CT-test scores are only correlated to belief argumentation.

## 7.2 Limitations

There are several limitations of the quantitative studies presented in Chapters 4, 5, and 6 that will be discussed here.

### 7.2.1 Coding

One limitation is coding the *argumentations* in only two gradations (i.e. naïve/inflexible and sophisticated). A finer graded analysis of the students' written texts might lead to a better understanding of their reasoning and developments in the pseudo-longitudinal parts of the study. However, during the development of the coding scheme, it proved to be hard to further distinguish the students' answers, as there is only a small number of students arguing in a sophisticated way. And even with only two types of argumentation, the studies provided significant and meaningful results. Nonetheless, in a follow-up study in Cologne (LeScMa – Learning the Science of Mathematics; Rott 2019, 2020; under review), the coding scheme was revised and now distinguishes four types of argumentation:

- *No justification*, when the text boxed remain empty or just contain one word like “yes” or “no”.
- *Naïve*, when a person does not respond meaningfully, for example saying “there might be miscalculations” as an argument for the uncertainty of mathematical knowledge.
- *Inflexible*, when the argumentation is meaningful, but does not contain arguments that substantially contribute to the argumentation like “mathematicians are humans and humans can err” or just repeat arguments from the initial statements.
- *Sophisticated*, when arguments given in the initial statements are combined in a new and meaningful way or arguments are used that do not stem from the initial statements.

This new coding scheme has been used with high interrater-agreement on almost 2000 written arguments (almost 1000 each for *certain vs. uncertain* as well as for *deductive vs. inductive*) by students. The new coding is mostly a way of further distinguishing the old “inflexible” code; the old “sophisticated” code has not been altered. Fittingly, the number of sophisticated answers in the new study is comparable to the ones coded in the sub-studies presented in this book.

This new way of coding in the Cologne study bears some similarities to the framework for teachers’ assessment of socioscientific argumentation by Christenson and Chang Rundgren (2014) (even though these authors’ study was only found after developing the new coding scheme). Christenson and Chang Rundgren base their framework on the SEE-SEP model[2] by Christenson et al. (2011) (see also Steffen and Hössle 2017). In the context of genetically modified organisms (GMO), Christenson and Chang Rundgren (2014, p. 3) differentiate between “*claims* (decision) and *justification* (with pros and cons)” – which are comparable to the different codes for *position* and *argumentation* in the study at hand. When a claim (e.g., “I am positive towards GMO”) is justified, the authors code for arguments for and against the claim in two dimensions: (1) value (grounded or non-grounded) and (2) based on content-knowledge. The latter, which is comparable to the argumentation codes from the LeScMa study, are coded with

[2] The SEE-SEP model links the six subject areas of sociology/culture (So), environment (En), economy (Ec), science (Sc), ethics/morality (Et) and policy (Po) with three aspects, knowledge, value and personal experience.

three different options: (A) incorrect content knowledge, (B) non-specific general knowledge, and (C) correct and relevant content knowledge included.

In the LeScMa study, students' beliefs were collected at three times, each time at the beginning of a winter term: (1) In October 2017 in lectures addressing mostly first semester students (ca. 600 participants), (2) in October 2018 in lectures addressing mostly third semester students (ca. 450 participants), and (3) in October (2019) in lectures addressing mostly fifth semester students (ca. 850 participants). With these data, pseudo-longitudinal as well as real longitudinal analyses (with almost 100 students who participated on all three surveys) can help to better understand the development of epistemological beliefs, which needs to be further studied (see Bromme et al. 2008).

The LeScMa study also expands the groups of participants from pre-service teachers to other university students with mathematics lectures in their programs of study (especially mathematics majors and business mathematics majors).

### 7.2.2 Writing

Another limitation of the quantitative studies is the effort that is needed to fill-in open-ended questionnaire items. In contrast to tick boxes in a closed (e.g., Likert scale) questionnaire, participants need to argue and need to write down their ideas, which can be both time-consuming and exhausting. Some participants might shy away from the trouble and skip answering items or argue with less effort and vigor than they could do. Gathering the data, this behavior could be observed as some participants ceased their efforts in answering the survey (containing of the CAEB, the CT test, and the open-ended belief questionnaire) after some time.

This effect can also be seen in the quantitative main study. The participants were asked to answer to the statements regarding the certainty of mathematical knowledge (see Chapter 5) first and then to the statements regarding the justification of mathematical knowledge (see Chapter 6). In the former, 463 surveys, and in the latter, only 439 surveys could be coded. The number of students' answers that were rated as sophisticated was 38 (8.2%) for certainty compared to 32 (7.3%) for justification. This small decrease indicates that most of the participants that are able to argue in a sophisticated way are also motivated to do so.

## 7.3 Analyses of Further Data

In the quantitative studies, some data has been obtained that has not yet been analyzed. For example, data about the participants' sex or their connotative beliefs (beyond what is described in Chapter 4; more on connotative beliefs in Sect. 7.4). Hypotheses regarding possible correlations between students' beliefs and their grades at school could be formulated and then evaluated. However, it is not intended to raise completely new research questions in this chapter of this book.

One additional question, however, is addressed here: As this book is the first publication in which both parts of the main study (Chapters 5 and 6) are presented together, it is of interest to investigate connections between both parts. Are the two dimensions of epistemological beliefs, certainty and justification of mathematical knowledge, independent from each other? This question is answered with a glance at the number of students that argue sophisticatedly. One could assume that students who are able to argue sophisticatedly in one dimension are able to do so in the other dimension as well.

In the dimension *certainty of mathematical knowledge*, the responses of 38 students (17 for certain and 21 for uncertain) were coded as sophisticated. In the dimension *justification of mathematical knowledge*, the responses of 32 students (13 for deductive and 19 for deductive) were coded as sophisticated. Looking at both data sets simultaneously reveals, however, that 65 students had at least one code for a sophisticated argumentation. In other words, only 5 students argued sophisticatedly in both dimensions.

A chi-square test (see Table 7.1) reveals that this number of students reasoning sophisticatedly in both parts of the quantitative main study is only slightly (and not significantly) larger than the expected number if both parts of the study were

**Table 7.1** Comparison of argumentation codes in both parts of the quantitative main study. The 24 participants that did not answer to the *deductive vs. inductive* part of the questionnaire were counted as "Inflexible"

| | | **Deductive / Inductive** | | |
|---|---|---|---|---|
| | | **Inflexible** | **Sophisticated** | **Sum** |
| **Certain / Uncertain** | **Inflexible** | 398 (395.6) | 27 (29.4) | **425** |
| | **Sophisticated** | 33 (35.4) | 5 (2.6) | **38** |
| | **Sum** | **407 + 24** | **32** | **463** |
| | | $\chi^2 = 1.56$ | df = 1 | $p = 0.212$ |

statistically independent from each other. This result is surprising insofar as reasoning should be an overarching competence; it indicates that the two dimensions of epistemological beliefs are not highly correlated.

Overall, those five students that argue sophisticatedly in both dimensions (see Table 7.2 for details) have CT scores as well as school grades that are significantly higher than the mean values of the total group. These findings should be further explored in the future.

## 7.4 Connotative Beliefs of Pre-Service Teachers

The importance of research on epistemological beliefs has been stated many times in this book. Surprisingly, so far, little research has been done into the question of the extent to which epistemological beliefs differ from individuals in terms of different subjects or disciplines. It is also still largely unclear to what extent teachers have different beliefs with regard to mathematics as a scientific discipline on the one hand and as a school subject on the other. Especially in mathematics, discontinuities between school and university are well known (cf. Ableitinger et al. 2013; Klein 1908). A differentiated view is important because such area-specific differences can be relevant for teacher behavior (Weinhuber et al. 2017).

One of the few studies on the possible difference between beliefs in mathematics as a scientific discipline or as a school subject comes from Beswick (2012), which is a case study with interviews and elaborate lesson observations. In this study, the beliefs and teaching behavior of two female teachers are examined. Beswick shows that teachers can have different beliefs between mathematics as a scientific discipline and as a school subject and that this affects their professional actions. She concludes that this issue requires further attention and research.

Therefore, with the data gathered in the quantitative pilot study (see Chapter 4), the goal is to examine the epistemological beliefs of pre-service mathematics teachers. In particular, it will be examined to what extent beliefs about mathematics as a scientific discipline and mathematics as a school subject can be distinguished economically with a closed questionnaire.

### 7.4.1 Methodology

One way to identify beliefs is to conduct interviews. This method has proven to be reliable, but very time-consuming both in its implementation and in its evaluation;

**Table 7.2** Compiled data for the five students with sophisticated argumentation in both parts of the quantitative main study

| Code Name | Sex | Study Program | University | Semester [1] | CT score | Abitur Grade [2] | Math Grade [3] | Positions |
|---|---|---|---|---|---|---|---|---|
| HDSV-25 | female | Upper Secondary | Essen | 3 / 3 | 1.939 | 1.0 | 14 | Certain, Deductive |
| MPAI-09 | male | Upper Secondary | Essen | 3 / 11 | 1.939 | 3.0 | – | Certain, Deductive |
| NMHR-20 | male | Upper Secondary | Essen | 3 / 5 | 0.710 | 1.8 | 15 | Uncertain, Deductive |
| KSEZ-15 | female | Primary | Freiburg | 5 / 5 | 1.939 | 1.8 | 13 | Certain, Deductive |
| SCAN-27 | female | Primary | Freiburg | 3 / 3 | 2.811 | 1.5 | 15 | Uncertain, Inductive |
| **Mean (all students)** | 71.9% f<br>28.1% m | 42.8% Primary<br>34.6% Lower Sec<br>22.7% Upper Sec | 59.8% FR<br>40.2% E | 3.03 / 3.76 | −0.129 | 2.3 | 10.1 | 47.5% Certain<br>52.5% Uncertain<br>34.9% Deductive<br>65.1% Inductive |

1 Number of semesters in the current course of studies / total number of semesters at a university
2 Ranging from 1.0 (best) to 4.0 (worst grade that passes the requirements for the Abitur, i.e. the university entrance degree)
3 Final math grade at Abitur, ranging from 15 (best) to 0 (worst)

the beliefs of larger samples cannot be collected with the aid of interviews for research-economic reasons.

In addition, the use of questionnaires (e.g., Grigutsch et al. 1998) is a widespread method of measuring beliefs, especially in German-speaking countries. Usually, these are questionnaires with closed items to record denotative beliefs: the questions should be answered consciously and reflected upon. Such questions are often about agreement on a Likert scale (disagree … agree) to statements like "Mathematical thinking is determined by abstraction and logic" (ibid., p. 42). Questionnaires with open items for recording denotative beliefs have so far only been used in isolated cases (e.g., Rott et al. 2015), presumably due to the high evaluation effort involved.

An alternative on which we will focus in the following is the use of a questionnaire to collect *connotative* beliefs: The respondents should not think about the items for a long time, but answer with an emotional emphasis. We used the CAEB (connotative aspects of epistemological beliefs, Stahl and Bromme 2007), which consists of 24 opposing pairs of adjectives such as "objective – subjective" or "completed – uncompleted". The test persons evaluate the character of a discipline on a 7-step scale for each pair of adjectives, which takes about 5 – 7 min in total. The instrument was validated by Stahl in two studies with more than 1000 subjects each. Among other things, it was shown that different disciplines such as genetics and physics are evaluated differently. Factor analyses were also used to identify two factors: *Texture* (beliefs about the structure and accuracy of knowledge) and *Variability* (beliefs about the stability and dynamics of knowledge).

### 7.4.2 The Study

147 students ($n = 105$ first semester students and $n = 42$ fourth semester students) at the PH Freiburg voluntarily participated in the study at the beginning of the 2013/14 winter semester. In addition to other data (e.g., denotative epistemological beliefs and critical thinking, see Rott et al. 2015, or Chapter 4, respectively), the participants were asked to complete the CAEB twice – with the two closely related disciplines of mathematics as a scientific discipline and mathematics as a school subject.

The data were analyzed using confirmatory factor analyses. The data were based on a measurement model developed on the basis of earlier studies in which the factors certainty/texture, simplicity and variability could be distinguished (Stahl and Bromme 2007; see also Chapter 4). Various fit indices were used to assess the model quality (Moosbrugger and Schermelleh-Engel 2007).

### 7.4.3 Results

In the analyses, a three-dimensional model with factors *safety/texture* as well as *simplicity* and *variability* proves to be acceptable ($\chi^2$/df = 1.56; RMSEA = 0.061; CFI = 0.886). A further differentiation of the two factors certainty and texture leads to a better model fit ($\chi^2$/df = 1.50; RMSEA = 0.058; CFI = 0.903), so that in the following, the comparison of the epistemological beliefs for mathematics as a scientific discipline vs. as a school subject is based on this model.

Figure 7.1 shows the extent to which mathematics was assessed differently as

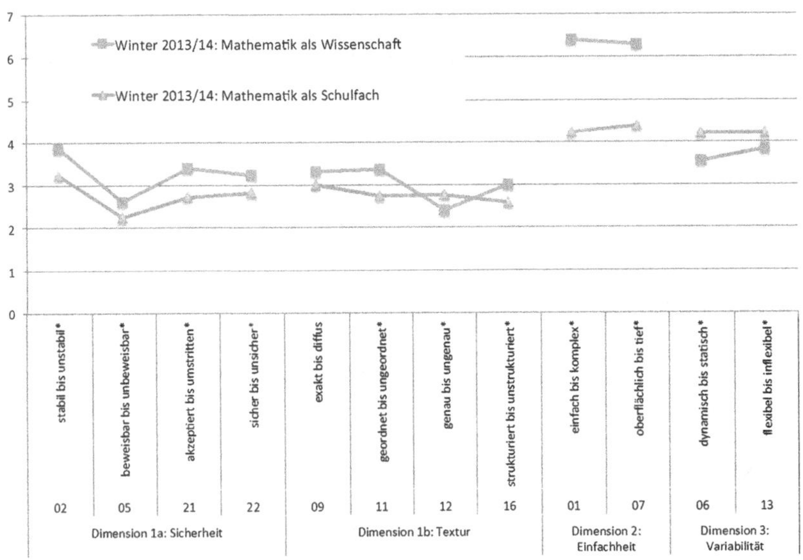


**Fig. 7.1** Mean values of CAEB items for the assessment of "mathematics as scientific discipline" and "mathematics as school subject". End poles of the response scale represent the verbal anchors represented on the horizontal axis. Significant differences (t-test for dependent samples, $p < 0.05$, two-sided testing) are highlighted with *

a scientific discipline / school subject based on the individual CAEB items: With the exception of Item 9, all differences are significant. The effect strengths for the items on the *simplicity* scale (dimension 2) are very large ($d_{01} = 1.7$; $d_{07} = 1.4$): Students perceive mathematics as a school subject much more simply and superficially than mathematics as a scientific discipline. For all other items, the effect strengths are rather small ($d < 0.50$). On the scales of *certainty* and *variability* (dimensions 1 and 3, resp.), there is a consistent trend that mathematics as a scientific discipline is perceived as less certain and more variable in comparison. In the assessment of *texture*, mathematics as a school subject is regarded as more ordered overall, but at the same time as somewhat less precise than mathematics as a scientific discipline (Item 12). If one also considers the number of semesters (4th semester dummycoded: $\chi^2/df = 1.44$; RMSEA = 0.055; CFI = 0.907), it can be seen that mathematics as a school subject seems to be perceived as significantly less certain ($\beta_{1a} = 0.55$) and more structured ($\beta_{1b} = 0.37$) as the course of study progresses, thus approaching the assessment of mathematics as a science. Conversely, advanced students rate mathematics as a scientific discipline significantly easier than students in the first semester ($\beta_{1a} = -0.70$).

### 7.4.4 Discussion

The results make it seem plausible that differences in the beliefs of pre-service mathematics teachers about mathematics as a scientific discipline and as a school subject can also be identified with the help of a comparatively inexpensive instrument such as the CAEB. This opens up numerous possibilities for studies in which the actions of teachers in connection with beliefs are to be investigated to carry out differentiated analyses. Before doing so, however, the approach should be confirmed by further validation studies.

## 7.5 Adapting the Methodology to Other Domains

The question remains whether the discussion presented in the chapters of this book is restricted to mathematics and its philosophy. Especially the certainty of mathematical knowledge – because of axiomatic reasoning and proofs – motivated the distinction between belief position and belief argumentation. So, can the ideas of this project be extended to other scientific domains, for example STEM (science, technology, engineering, and mathematics) fields? Especially adapting the methodology, using the open-ended questionnaire with opposite quotes that the participants should refer to, should be questioned.

For example, in STEM education, there is branch of research dealing with epistemological beliefs, called research on the "nature of science" (NoS) (e.g., Höttecke 2001; Lederman 2007; see also Matthews 2012, for historical background). NoS is dedicated to questions regarding the philosophy of natural sciences like the epistemology and ontology of knowledge in disciplines like chemistry and physics. In this context, the role of experiments, models, or theories in the creation or discovery of knowledge is specifically addressed. In NoS, researchers also deal with learners' (developments of) beliefs about natural sciences and ways of measuring such beliefs.[3]

Altogether, there are many parallels and similarities between research on epistemological beliefs in psychology and research on beliefs in NoS; however, there are also differences. Neumann and Kremer (2013) compared the two concepts and identified five aspects to structure the discussion about both research traditions. Those aspects are (1) discipline specificity; (2) content; (3) perspective of the first vs. third person; (4) knowledge vs. views; and (5) normative vs. descriptive approaches. These aspects are now addressed briefly.

Regarding (1), NoS exclusively refers to natural sciences, whereas in psychological research, it has not yet been clarified whether epistemological beliefs are a general concept or whether they are domain-specific. Addressing the (2) content, there is not such a clear factor structure in NoS compared to epistemological beliefs with factors like "certainty of knowledge" or "source of knowledge". This may be due to the lack of large factor-analytical studies in STEM education or a more normative approach of NoS. Regarding the (3) perspective, psychology deals with individual beliefs of learners, while NoS is more about general beliefs about learning and acting in natural sciences. (4) NoS is more about (conscious) knowledge, while epistemological beliefs are often conceptualized in a more unconscious way. Closely related to the previous aspect are (5) the approaches to conceptualizing beliefs in the two research traditions. In NoS, the debate centers on normative knowledge, something to be taught in sciences courses and classes, whereas epistemological beliefs are more descriptive (even if this also indicates a direction for learning goals with categories such as "naïve / sophisticated"). (Neumann & Kremer 2013).

[3] In contrast to psychological research on epistemological beliefs, researchers in NoS more often use instruments that are not self-report, Likert scale questionnaires. For example, the open-ended VNOS (views of nature of science) questionnaire by Lederman et al. (2002) and/or the use of interviews (e.g., Höttecke & Rieß, 2007).

As mentioned before, the NoS debate is to some degree normative, which is possible, because "little disagreement exists among philosophers, historians, and science educations" (Lederman 2007, p. 833) regarding its core contents. These contents can be summarized as follows (often called the "Lederman Seven"):

> Among the characteristics of scientific knowledge corresponding to this level of generality are that scientific knowledge is tentative (subject to change), empirically based (based on and/or derived from observations of the natural world), and subjective (involves personal background, biases, and/or is theory-laden); necessarily involves human inference, imagination, and creativity (involves the invention of explanations); and is socially and culturally embedded. Two additional important aspects are the distinction between observations and inferences, and the functions of and relationships between scientific theories and laws. (Lederman, 2007, p. 833)

Thus, for example, it is generally accepted that scientific knowledge is tentative. A famous historical example is the replacement of Newton's law of universal gravitation by Einstein's theory of general relativity. Against this backdrop, it should be hard to nearly impossible to adapt the questionnaire to other disciplines than mathematics, especially to science education, as there are no different positions that are equally correct and that can be argued for or against; for example, no expert would say that physical knowledge is certain. The same is true for inductive/ deductive reasons, because in natural sciences, there is no deductive way of reasoning. Therefore, there would be no independent distinction between *position* and *argumentation* in an adaptation of the LeScEd questionnaire.

However, some authors point out the danger that those core contents – or tenets – of NoS "may be easily misinterpreted and abused" (Clough 2007, p. 3), for example as declarative knowledge to be memorized.

> For instance, when addressing the historical tentative character of science years ago while teaching high school science, my students would jump from the one extreme of seeing science as absolutely true knowledge to the other extreme as unreliable knowledge. Extensive effort was required to move them to a more middle ground position. Colleagues have told me of students who have asked why they have to learn science content if it's always changing. The same was true of issues regarding invention/discover, subjectivity/objectivity, private/public science, and scientific methodology. (Clough, 2007, p. 3)

Clough states that it is correct that scientific knowledge is tentative, especially from a historical perspective; but then he relativizes: "However, the tenet ignores the durable character of well-supported scientific knowledge. Students who claim that science is tentative without acknowledging the durability of well-supported scientific knowledge can hardly be said to understand the nature of science." (ibid., p. 2) Similarly, Matthews (2012) points out that Lederman's "NOS checklist […] appears on classroom walls somewhat like the Seven NOS Commandments [which can] function as a mantra, as a catechism, as yet another something to be learnt [without] teachers and students reading, analysing, and coming to their own views about NOS matters" (p. 3).

With this debate in mind, it should be possible to confront students with two opposing statements, for example regarding the certainty, stability or durability of scientific knowledge and have them give reasons for their choice. A first attempt at adapting the LeScEd questionnaire is made in chemistry education (Euskirchen 2019; a Master thesis that was co-supervized by the autor (BR)).

There is another aspect that this comparison of NoS with epistemological beliefs in mathematics education or mathematical views, respectively, can contribute to. In mathematics education, a systematic acquisition of knowledge about epistemology has so far rarely been discussed. In contrast to other STEM didactics and the NoS debate, there is no broad discussion about the development of epistemological beliefs in mathematics education, neither is research on this topic. Additionally, in NoS, there are various attempts at systematically integrating epistemology into teaching concepts (e.g., Akerson et al. 2000) and ways of evaluating them accordingly (ibid.; Holbrook & Rannikmae 2009). For mathematics, similar courses for the development of sophisticated epistemological beliefs or a critically reflected "Nature of Mathematics" are generally not integral components of education. In teacher training programs, one seems to assume that adequate ideas and beliefs are formed solely through intensive study of "mathematics from an advanced standpoint" (CMBS 2010).

## References

Ableitinger, C., Kramer, J., & Prediger, S. (Eds.). (2013). *Zur doppelten Diskontinuität in der Gymnasiallehrerbildung – Ansätze zu Verknüpfungen der fachinhaltlichen Ausbildung mit schulischen Vorerfahrungen und Erfordernissen.* Springer.

Akerson, V. L., Abd-El-Khalick, F., & Lederman, N. G. (2000). Influence of a Reflective Explicit Activity-Based Approach on Elementary Teacher's Conceptions of Nature of Science. *Journal of Research in Science Teaching, 37*(4), 295–317

Beswick, K. (2012). Teachers' beliefs about school mathematics and mathematicians' mathematics and their relationship to practice. *Educational Studies in Mathematics, 79*, 127–147

Bromme, R., Kienhues, D., & Stahl, E. (2008). Knowledge and Epistemological Beliefs: An Intimate but Complicate Relationship. In M. S. Khine (Ed.), *Knowing, Knowledge and Beliefs: Epistemological Studies across Diverse Cultures.* (pp. 423–441). Springer.

Christenson, N., Chang Rundgren, S.-N., & Höglund, H.-O. (2011). Using the SEE-SEP model to analyze upper secondary students' use of supporting reasons in arguing socioscientific issues. *Journal of Science Education and Technology, 21*, 342–352

Christenson, N., & Chang Rundgren, S.-N. (2014). A framework for teachers' assessment of socio-scientific argumentation: An example using the GMO issue. *Journal of Biological Education, 49*(2), 204–212

Clough, M. P. (2007) Teaching the Nature of Science in Secondary and Post-Secondary Students: Questions Rather Than Tenets. *The Pantaneto Forum, 25.*

CMBS: Conference Board on the Mathematical Sciences. (2010). *The Mathematical Education of Teachers II*. American Mathematical Society.

Euskirchen, M. (2019). *Konzeption eines Fragebogens zur Erhebung von epistemologischen Überzeugungen bei Lehrenden im Fach Chemie und dessen Pilotierung.* Master Thesis, University of Cologne.

Grigutsch, S., Raatz, U., & Törner, G. (1998). Einstellungen gegenüber Mathematik bei Mathematiklehrern. *Journal Für Mathematik-Didaktik, 19*(1), 3–45

Hofer, B. K., & Pintrich, P. R. (1997). The development of epistemological theories: Beliefs about knowledge and knowing and their relation to learning. *Review of Educational Research, 67*(1), 88–140

Holbrook, J., & Rannikmae, M. (2009). The Meaning of Scientific Literacy. *International Journal of Environment & Science Education, 4*(3), 275–288

Höttecke, D. (2001). Die Vorstellungen von Schülern und Schülerinnen von der „Natur der Naturwissenschaften". *Zeitschrift Für Didaktik Der Naturwissenschaften, 7*, 7–23

Höttecke, D. & Rieß, F. (2007). How Do Physics Teacher Students Understand the Nature of Science? An Explorative Study of a Well Informed Investigational Group. *Paper presented at the Ninth International History, Philosophy, Sociology & Science Teaching Conference* (IHPST), Calgary, Canada 2007, June 28–31, 2007, www.ucalgary.ca/ihpst07/proceedings/IHPST07%20papers/124%20hoettecke.pdf

Kahneman, D. (2011). *Thinking, fast and slow.* Penguin Books Ltd.

Klein, F. (1908). *Elementarmathematik vom höheren Standpunkte aus.* B. G. Teubner.

Lederman, N. G., Abd-El-Khalick, F., Bell, R. L., & Schwartz, R. S. (2002). Views of nature of science questionnaire: Toward valid and meaningful assessment of learners' conceptions of nature of science. *Journal of Research in Science Teaching, 39*(6), 497–521

Lederman, N. G. (2007). Nature of Science: Past, Present, and Future. In S. K. Abell & N. G. Lederman (Eds.), *Handbook of research on science education.* (pp. 831–879). Routledge.

Matthews, M. R. (2012). Changing the focus: From nature of science (NOS) to features of science (FOS). In M. S. Khine (Ed.), *Advances in Nature of Science Research.* (pp. 3–26). Springer.

Moosbrugger, H., & Schermelleh-Engel, K. (2007). Exploratorische (EFA) und Konfirmatorische Faktorenanalyse (CFA). In H. Moosbrugger & A. Kelava (Eds.), *Testtheorie und Fragebogenkonstruktion.* (pp. 307–324). Springer.

Neumann, I., & Kremer, K. (2013). Nature of Science und epistemologische Überzeugungen – Ähnlichkeiten und Unterschiede. *Zeitschrift Für Didaktik Der Naturwissenschaften, 19*, 209–232

Rott, B. (2019). Epistemologische Überzeugungen zur Mathematik von Studierenden des Fachs – Einblicke in ein Forschungsprojekt. *DMV-Nachrichten, 27*(1), 13–17

Rott, B. (2020, online first). Inductive and deductive justification of knowledge: epistemological beliefs and critical thinking at the beginning of studying mathematics. Educational Studies in Mathematics. https://link.springer.com/article/10.1007/s10649-020-10004-1

Rott, B., Leuders, T., & Stahl, E. (2015). Assessment of mathematical competencies and epistemic cognition of pre-service teachers. *Zeitschrift Für Psychologie, 223*(1), 39–46

Rott, B., Groß Ophoff, J., & Leuders, T. (2017). Erfassung der konnotativen Überzeugungen von Lehramtsstudierenden zur Mathematik als Wissenschaft und als Schulfach. In U. Kortenkamp & A. Kuzle (Eds.), *Beiträge zum Mathematikunterricht 2017.* (pp. 1101–1104). WTM.

Stahl, E., & Bromme, R. (2007). The CAEB: An instrument for measuring connotative aspects of epistemological beliefs. *Learning and Instruction, 17*, 773–785

Stahl, E. (2011). The Generative Nature of Epistemological Judgments: Focusing on Interactions Instead of Elements to Understand the Relationship between Epistemological Beliefs and Cognitive Flexibility (Chapter 3). In J. Elen, E. Stahl, R. Bromme, & G. Clarebout (Eds.), *Links Between Beliefs and Cognitive Flexibility – Lessons Learned.* (pp. 37–60). Springer.

Steffen, B., & Hössle, C. (2017). Kriteriengeleitete Diagnose von Bewertungskompetenz im Fach Biologie – Ein Blick in den internationalen Raum. *MNU Journal, 3*, 201–207

Weinhuber, M., Lachner, A., Leuders, T., & Nückles, M. (2017). *Context affects teachers' principle-orientation of explanations.* Symposium held at the 17th Biennial Conference of the European Association for Research on Learning and Instruction (EARLI), Tampere, Finland.

Printed in the United States
by Baker & Taylor Publisher Services